Durability Analysis of Aluminized Type 2 Corrugated Metal Pipe

1. Report No. FHWA-RD-97-140	2. Government Accession No.	3. Recipient's Catalog No.		
4. Title and Subtitle DURABILITY ANALYSIS OF ALUMINIZED TYPE 2 CORRUGATED METAL PIPE		5. Report Date		
		6. Performing Organization Code		
7. Author(s) J. Peter Ault, P.E., and James A. Ellor, P.E.		8. Performing Organization Report No.		
9. Performing Organization Name and Address Ocean City Research Corp. Tennessee Ave & Beach Thorofare Ocean City, NJ 08226		10. Work Unit No. (TRAIS)		
		11. Contract or Grant No. DTFH61-94-C-00213		
12. Sponsoring Agency Name and Address Office of Infrastructure Research and Development Federal Highway Administration 6300 Georgetown Pike McLean, VA 22101-2296		13. Type of Report and Period Covered Final Report Oct. 1994 – Dec. 1996		
		14. Sponsoring Agency Code		
15. Supplementary Notes Contracting Officer's Technical Representative (COTR): John O'Fallon, HRDI-07 Technical Assistance: Anthony Welch, HFPD-8, and Philip Thompson, HIBT-20				

16. Abstract

The Literature Review and Field Studies portion of this investigation were completed by August 1995. Both revealed myriad factors affecting culvert durability. The literature review considered more than 140 research papers on or relating to the topic of culvert durability. Of these papers, roughly 60 were considered further and are included in an annotated bibliography. Many references are made in the report to past research efforts and conclusions from these papers. Summaries of these conclusions include laboratory and field research on various culvert materials and durability methods. These summaries are included to help in both the selection of pipe material and in the use of durability prediction methods. Those papers presenting field data similar to that used in the present study (including pit depth, age, and original thickness) were consolidated into a database of information on more than 240 galvanized culverts.

The focus of field studies conducted during this investigation centered on the performance of Aluminized Type 2 culverts. Of the 32 culverts inspected during this investigation, 21 culverts were part of a previous Federal Highway Administration (FHWA) study. These culverts were located in Alabama and Oregon. Eight culverts at five sites in Maine were added with the expansion of these field studies. In situ field measurements of pH and resistivity were taken at each culvert. The abrasive situation at each culvert was characterized, including slope, flow velocity, and abrasive material. Each culvert was photo-documented and many were videotaped for future reference. Coupon samples were taken at each location for pit-depth analysis. Soil samples were removed for laboratory resistivity measurements.

Conclusions from past research, the database, and present field studies have been used to evaluate current durability prediction methods. Advantages and disadvantages of various culvert materials are discussed, with correlations drawn from the literature review and field studies.

17. Key Words Corrugated metal pipe, corrosion, aluminum-coated, durability prediction, pit depth, culvert		18. Distribution Statement No restrictions. This document is available to the public through the National Technical Information Service, Springfield, VA 22161.		
19. Security Classif. (of this report) Unclassified	20. Security Classif. (of this page) Unclassified	21. No. of Pages 106	22. Price	

Form DOT F 1700.7   (8-72)   Reproduction of this page authorized

SI* (MODERN METRIC) CONVERSION FACTORS

APPROXIMATE CONVERSIONS TO SI UNITS

Symbol	When You Know	Multiply By	To Find	Symbol
LENGTH				
in	inches	25.4	millimeters	mm
ft	feet	0.305	meters	m
yd	yards	0.914	meters	m
mi	miles	1.61	kilometers	km
AREA				
in^2	square inches	645.2	square millimeters	mm^2
ft^2	square feet	0.093	square meters	m^2
yd^2	square yards	0.836	square meters	m^2
ac	acres	0.405	hectares	ha
mi^2	square miles	2.59	square kilometers	km^2
VOLUME				
fl oz	fluid ounces	29.57	milliliters	mL
gal	gallons	3.785	liters	L
ft^3	cubic feet	0.028	cubic meters	m^3
yd^3	cubic yards	0.765	cubic meters	m^3

NOTE: Volumes greater than 1000 l shall be shown in m^3.

Symbol	When You Know	Multiply By	To Find	Symbol
MASS				
oz	ounces	28.35	grams	g
lb	pounds	0.454	kilograms	kg
T	short tons (2000 lb)	0.907	megagrams (or "metric ton")	Mg (or "t")
TEMPERATURE (exact)				
°F	Fahrenheit temperature	5(F-32)/9 or (F-32)/1.8	Celsius temperature	°C
ILLUMINATION				
fc	foot-candles	10.76	lux	lx
fl	foot-Lamberts	3.426	candela/m^2	cd/m^2
FORCE and PRESSURE or STRESS				
lbf	poundforce	4.45	newtons	N
lbf/in^2	poundforce per square inch	6.89	kilopascals	kPa

APPROXIMATE CONVERSIONS FROM SI UNITS

Symbol	When You Know	Multiply By	To Find	Symbol
LENGTH				
mm	millimeters	0.039	inches	in
m	meters	3.28	feet	ft
m	meters	1.09	yards	yd
km	kilometers	0.621	miles	mi
AREA				
mm^2	square millimeters	0.0016	square inches	in^2
m^2	square meters	10.764	square feet	ft^2
m^2	square meters	1.195	square yards	yd^2
ha	hectares	2.47	acres	ac
km^2	square kilometers	0.386	square miles	mi^2
VOLUME				
mL	milliliters	0.034	fluid ounces	fl oz
L	liters	0.264	gallons	gal
m^3	cubic meters	35.71	cubic feet	ft^3
m^3	cubic meters	1.307	cubic yards	yd^3
MASS				
g	grams	0.035	ounces	oz
kg	kilograms	2.202	pounds	lb
Mg	megagrams (or "metric ton")	1.103	short tons (2000 lb)	T
TEMPERATURE (exact)				
°C	Celsius temperature	1.8C + 32	Fahrenheit temperature	°F
ILLUMINATION				
lx	lux	0.0929	foot-candles	fc
fl	candela/m^2	0.2919	foot-Lamberts	fl
FORCE and PRESSURE or STRESS				
N	newtons	0.225	poundforce	lbf
kPa	kilopascals	0.145	poundforce per square inch	lbf/in^2

TABLE OF CONTENTS

LIST OF FIGURES

LIST OF FIGURES (continued)

LIST OF TABLES

INTRODUCTION

Durability of Culverts and Special Coatings is intended to provide up-to-date storm drain and culvert material selection methods and techniques to assist the highway engineer in designing culverts. While the program was to consider a variety of materials, the predominant focus was on Aluminized Type 2 corrugated steel pipe (CSP) and its performance versus galvanized CSP. This report is divided into two main sections—field investigation and literature review. The field investigation presents the results of a field evaluation of 32 pipes and their performance relative to that reported in FHWA-FLP-91-006, *Durability of Special Coatings for Corrugated Steel Pipe*. The present report also includes a review of previous research areas to evaluate conclusions and the utility of existing culvert life prediction methods. The literature review presents a discussion of the various factors affecting culvert durability and its prediction.

Field studies were conducted during this investigation to update and expand those reported in FHWA-FLP-91-006, *Durability of Special Coatings for Corrugated Steel Pipe*. The focus of the field studies was on Aluminized Type 2; however, field evaluations of alternative materials and coatings are included. In total, 21 Aluminized Type 2 pipes were investigated for analysis and comparison to previous data. Three additional Aluminized Type 2 pipes and eight other corrugated metal pipes were inspected because of their proximity to the pipes in this study. Their condition is reported for reference only.

A comprehensive literature review was performed to explore past and present research areas concerning culverts. A discussion of alternative materials, including aluminum alloy, Galvalume, galvanized steel, and bituminous-coated galvanized steel, is presented in this report. Also discussed in this report are various methodologies for the prediction of culvert durability and the methods used to select culvert materials. Consideration is given to methods developed by the American Iron and Steel Institute (AISI), National Corrugated Steel Pipe Association (NCSPA), California, Colorado, Florida, New York, and Utah. The strengths and weaknesses of each method as durability prediction tools are discussed. Where possible, correlations were drawn between durability predictions and observations made during the field studies in this report.

Prediction of culvert durability is difficult because of the number of variables in natural environments that affect culvert service life. Many variables are easy to measure (e.g., pH, resistivity, water chemistry) while others are difficult (e.g., abrasion). However, even easy-to-measure variables will have spatial and temporal variations (e.g., they will not be a single value over the length and life of a culvert). This complexity in exposure environment must be considered for durability predictions to be of practical use.

A variety of methods are available to predict the durability of culverts. The most commonly used is California Test Method No. 643 (California Method) developed by Caltrans (California Department of Transportation). Some references state that the California Method is too liberal while others say that it is too conservative. There is much field data contained in the literature review to suggest that it may be either, depending on other environmental conditions. Many arguments for and against the California Method stem from disagreements over the most important factors in culvert deterioration. The California Method is based on hydrogen-ion concentration (pH) and minimum resistivity of the soil or the water (though other factors are considered when resistivity is below 1000 ohm-cm). Additional modifications correct for dissolved salt concentrations and free carbon dioxide content. Different durability prediction methods will be discussed in this report.

One environmental factor that does not seem to receive enough attention is abrasion. Abrasion is considered in the California Method, but only to the extent that the 7,000 culverts forming the database for that method are from a cross-section of abrasive environments. This may be a factor in the accuracy ($\pm$12 years) of the California Method. The current study evaluates the abrasion potential of the environment by measuring the slope of the installation, the velocity of the flow, the size of any abrasive material found in the culvert, and by noting other culvert characteristics that suggest the flow experienced by the installation.

Finally, there is a need for improved ways to define failure modes and mechanisms for concrete and plastic culvert materials so that their lives can be better predicted for quantitative comparison to the steel culverts and coatings that receive so much attention.

CONCLUSIONS

1. The data gathered in this study provides a limited basis on which to determine the advantages of Aluminized Type 2 versus galvanized steel culverts. Only 21 Aluminized Type 2 culverts formed the basis for this study. In contrast, the California Method was based on 7,000 culverts.

2. The Aluminized Type 2 pipes inspected are performing similarly to those reported in the 1990 study, after adjustment for the additional exposure time. The Aluminized Type 2 pipes that were perforated at the time of the initial study are still functioning adequately.

3. A variety of service-life multipliers may be justified for Aluminized Type 2 pipe versus the California Method predictions for galvanized pipe. Our field studies suggest that in the absence of abrasion, an Aluminized Type 2 pipe may have a service life up to eight times that predicted for galvanized pipe by the California Method. (If only waterside corrosion is considered, the ratio becomes 3.5.) This is a very similar result to the previous study by Potter et al., which suggested that Aluminized Type 2 should last 6.2 times the life predicted for galvanized culvert. In these studies, galvanized culverts were not exposed in parallel; thus it is difficult to determine the accuracy of the California Method in identical environments.

4. All of the data and observations in this study suggest that Aluminized Type 2 will perform as well as or better than galvanized pipe. The actual service-life multiple for a given pipe will vary depending on the specific environment. Under extreme conditions, it is expected that the materials will perform in a relatively similar manner (i.e., last a long time or fail rapidly).

5. For the Aluminized Type 2 CSP in the study, there are at least two pitting tendencies. At the higher pitting rates, the major influence of the environmental variables observed is the severity of the bed load. At the low pitting rate, the pitting appears linear with time. Determining the pitting tendency is essential to predicting the eventual life of CSP.

6. There is not a consistent and comprehensive methodology to inventory, inspect, and evaluate the culverts in the field (similar to the biennial bridge inspection procedures) among all States. If such a methodology were developed (and used), the resulting database would provide much useful information for the development of improved service-life predictions.

7. The California Method is intended to provide guidance in selecting culvert materials and thicknesses. It is used by several agencies and has become widely accepted. However, users should remember that the method has an estimated accuracy of ± 12 years. Furthermore, the predictions do not change for varying abrasive conditions. The accuracy can be improved if the practical experience of the design agency is drawn on to "calibrate" the California Method to their applications and specific environments.

RECOMMENDATIONS

1. At a minimum, incorporate the existing Federal Lands Highway Design Guidance abrasion rating system (Levels 1 through 4) into culvert condition assessment and durability prediction practices.

2. Develop an inspection protocol similar to that used for biennial bridge inspections to inventory and inspect the Nation's culverts. Centralize the data generated to develop a decision-making algorithm for the selection of alternative culvert materials.

3. Evaluate alternative engineering techniques to better define environmental conditions, including abrasion, peak flow, and nominal flow. Techniques might be obtained from studies of river-scour or riverbed hydrology. These techniques might be used to improve the existing Federal Highway Administration (FHWA) abrasion rating system. The problem with the present classification scheme is that water velocity and bed loads are not independent factors. For example, there is no provision for high-velocity flow with minimal or no bed load. High velocity without a bed load would be expected to be less abrasive than lower velocity conditions with a bed load. A standard laboratory test method to investigate bed load/flow relationships would be the NCSPA protocol, "Evaluation Methodology for Corrugated Steel Pipe Coating/Invert Treatments."

4. Conduct a comprehensive national study to determine the service life of culverts and to develop an improved prediction model. Further attempts to provide durability models for different types of culvert in different environments must consider a wider range of parameters than those considered herein or those considered by most industry models. Analysis of the data should use extreme value methods.

FIELD INVESTIGATIONS

Approach and Methodology

One task of this investigation was to update and expand upon the field studies listed in table 4 of Report No. FHWA-FLP-91-006. The sites listed in this table are from the Santiam Highway in Oregon and the Natchez Trace Parkway in Alabama. The focus of study at these two locations is on Aluminized Type 2 coating over steel. There are a few pipes at these locations with other coatings, thereby permitting limited cross-comparison.

These studies were expanded to include five other Aluminized Type 2 coated pipes. Many sites uncovered in the literature review were considered. Five culverts installed by the Maine Department of Transportation (DOT) as part of its ongoing culvert material study were selected for this study. A limited analysis of some of the Maine sites was included in the previous study. These sites are especially interesting because two out of the five are tandem installations of more than one type of material. As a result, it was possible to investigate galvanized and Galvalume materials in the same installation as the Aluminized Type 2. This type of installation provided for a convenient cross-comparison of materials subject to the same environmental conditions.

In all, 32 culverts at 26 installations in 3 different States were considered. Twenty-four of the pipes were Aluminized Type 2 coated. The following outlines the field inspection procedure:

1. Pipe Location - The pipe's location was positively determined using information provided by previous researchers, State DOT's, and others. The pipe's State, county, road, mile marker/station number, and other information used to locate the site were also documented. Maps showing the locations were included. Photographs were taken of the general area to show the general terrain of each site.

2. Pipe Description - The pipe's corrugation type, diameter, length, slope, installation date, and any other pertinent pipe characteristics discovered were documented during the field studies.

3. Soil - The site's soil temperature, resistivity, pH, and abrasiveness were documented. Numerous resistivity measurements were made in situ with a Collins Rod. These measurements were made in the stream bed, around the pipe end, and on the inside of the pipe where coupons were removed. A sample of the soil was collected from the site for minimum soil resistivity measurements using the soil box. Triplicate measurements of the soil pH were taken from various locations in the stream bed. The abrasive potential of the pipe was evaluated at each site, including the size of rocks and pebbles found in or around the pipe at the time of the investigation.

4.   Water - The water's temperature, resistivity, pH, chemistry, and flow were investigated. Both resistivity and pH were measured with portable meters at various locations in and around the pipe. Water chemistry measurements were performed using portable test kits. Water chemistry measurements included chlorides, hardness, carbon dioxide, and alkalinity. The flow description at the time of the study was characterized. Any clues about flow conditions at other times were noted (i.e., high-flow stains, rocks in the invert, wear marks). Flow rates were estimated by a float in the water stream. Water that did not fill the corrugations was considered standing water. A description of the most recent rain activity was documented. At each site the abrasion potential was characterized in accordance with the Federal Lands Highway Design Guidance as follows:

Level 1 - **Non-abrasive** conditions exist in areas of no bed load and very low flow velocity. This is also the condition assumed for the soil side of drainage pipes.

Level 2 - **Low abrasive** conditions exist in areas of minor bed load of sand and velocities of 1.5 m/s (5 ft/s) or less.

Level 3 - **Moderately abrasive** conditions exist in areas of moderate bed load of sand and gravel and velocities between 1.5 m/s and 4.5 m/s (5 and 15 ft/s).

Level 4 - **Severely abrasive** conditions exist in areas of heavy bed load of sand, gravel, and rock and velocities exceeding 4.5 m/s (15 ft/s).

5.   Pipe Samples - Coupon samples 3.8 cm (1.5 in) in diameter were removed from the pipe at stations on the pipe circumference of 6 o'clock, water line, and 12 o'clock. These coupons were used to positively identify the coating and metallographically determine the remaining coating thickness.

6.   Pipe Interior - Each pipe's interior was visually inspected. The amount and type of corrosion products on the pipe were characterized. Photographs and videotape were made for future reference.

Each pipe's physical condition was documented photographically. Videotape recordings were made of the Oregon and Maine sites to document flow, deterioration, environment, and general observations.

Further sample analysis was performed in the laboratory. This included soil-resistivity measurements, pit-depth evaluation using a Mitutoyo digital micrometer, and metallographic analysis. The metallographic analysis included cutting and sectioning the core samples. Once sectioned, they were mounted for analysis of coating thickness and deterioration under the microscope. Coating material was also positively identified using chemical procedures in the Department of Defense handbook, *Rapid On-Site Identification of Metals and Alloys* (Report No. DOD-HDBK-249). These tests involved chemical spot-testing to verify metallic alloys.

Analysis

Table 1 presents a summary of the field data gathered during this investigation from Alabama, Oregon, and Maine. A total of 32 pipes (including 24 Aluminized Type 2, 3 Galvalume, 3 bituminous-coated galvanized, 1 aluminum alloy, and 1 galvanized) were inspected during this investigation. This study has concentrated on field evaluation of 21 Aluminized Type 2 pipes. Three extra Aluminized Type 2 culverts and a few pipes of other materials were looked at because of their easy access while evaluating the Aluminized Type 2. A list of all pipes inspected is contained in Appendix D.

Table 1 lists the pipes grouped by coating. The coating alloy was confirmed using chemical test procedures. The coupon thickness and pit-depth measurements were made with a digital Mitutoyo micrometer. Field observations during this work and that of others found that most corrosion occurs in the invert portion of the pipe between the 5 o'clock and 7 o'clock positions. In this area, the worst corrosion seemed to occur on the crests of the corrugations. Though it does not always occur, there have been documented cases where the increased oxygenation at the water line can cause more corrosion than the lowest portion of the submerged invert. For this reason, deepest pit measurements were made on both the invert (6 o'clock) and water-line coupon samples. The culverts in this study all exhibited maximum corrosion at the invert. Therefore, the deepest pit measurements taken from these samples were used to calculate the "actual" perforation for comparison with the life predicted by the California Test Method 643 (California Method). To obtain the original thickness, a series of random measurements were made on each crown sample and then averaged. For the purpose of making these calculations, it is assumed that the crown (12 o'clock position) of the pipe has experienced a negligible amount of corrosion. The "Percent Perforation" on this chart represents the depth of the deepest pit as a percentage of the original thickness.

The soil and water pH and resistivities in the table are from measurements made in the field. As in Report No. FHWA-FLP-91-006, the actual percent perforation of the pipe is compared with predictions made by the California Method. The California Method does not make predictions for Aluminized Type 2. The previous study utilized comparisons of Aluminized Type 2 against the *prediction* for galvanized as a means of quantifying the performance of Aluminized Type 2. The analysis in this study follows that of the previous study in order to update the findings.

The heading "Calif. Pred. Life Years" in table 1 represents the number of years to first perforation predicted by the California Method for the gauges and environmental data listed. In accordance with the California Method, minimum resistivity and pH are used in the calculations. For this study, we are using the lowest age predicted by either the soil-side or water-side pH and resistivity. In most cases this is the soil side. The column headed "Actual Age - Percent of Calif. Pred." gives the percent of the California Method predicted life that the culvert has been in service. Previous investigators have suggested that the percent perforation of the maximum pit should match the percent of the California Method predicted life and that any deviation suggests a correspondingly longer or shorter service life. The current age of each pipe was taken from best available documentation.

Table 1. Summary of Field Data on Aluminized Type 2 Pipes
Natchez Trace(NT), Santiam Highway(SH), and Maine(ME) Sites

Culvert Location	Pipe Gage	Slope Deg.	Velocity fps	Bedload	Thickness - inches			Percent Perforation	Soil		Water		Calif. Pred. Life Years	Actual Age	
					Crown	Invert	Waterline		pH	Resistivity	pH	Resistivity		Years	Percent of Calif. Pred.
ME Dexter	16	0	8	heavy	0.057	0.011	0.049	80.70%	6.7	5138	7.4	7389	30	16	53%
ME Garland	14	3	1	minor	0.072	0.070	0.072	2.78%	6.9	8130	6.8	9147	44	10	23%
ME New Gloucester	14	2	3	minor	0.072	0.071	0.073	1.39%	5.3		6.4	7426	38	16	42%
ME Orrington	14	0	1	minor	0.076	0.070	0.071	7.89%	6.5	3609	6.9	6186	32	10	31%
ME Ripley	12	1	2	minor	0.098	0.097	0.097	1.02%	5.4	5036	6.6	9677	36	16	44%
NT 310.0	16	4	4	moderate	0.058	0.000	0.054	100.00%	6.8	4710	7.0	17241	30	14	47%
NT 310.1	16	1	3	moderate	0.056	0.000	0.056	100.00%	5.8	2646	7.1	22556	17	14	82%
NT 310.6 North	16	1	*	minor	0.057	0.054	0.056	5.26%	6.8	6148	7.2	5272	32	14	44%
NT 311.9 South	16	3	*	minor	0.056	0.056	0.056	0.00%	7.6	1601	7.7	6061	38	14	37%
NT 312.4 East	16	0	0	none	0.057	0.056	0.057	1.75%	7.0	2594	7.2	2609	29	14	48%
NT 312.4 Single	16	0	*	none	0.057	0.057	dry	0.00%	7.0	2882	dry	dry	30	14	47%
SH 100+15	16	5	4	none	0.058	0.056	0.058	3.45%	5.3	3258	7.5	23438	17	14	82%
SH 104+45 East	10	3	4	none	0.126	0.124	0.123	1.59%	5.7	4631	7.2	19608	48	14	29%
SH 104+45 West	10	3	5	moderate	0.130	0.129	0.128	0.77%	5.7	4631	7.2	19608	48	14	29%
SH 123+76	16	3	2	none	0.056	0.055	0.055	1.79%	5.6	11152	6.6	18868	27	14	52%
SH 13+00	16	3	1	minor	0.057	0.057	0.057	0.00%	6.7	3464	6.8	46875	26	14	54%
SH 18+20	16	4	1	none	0.057	0.054	0.056	5.26%	6.5	10352	7.3	18987	32	14	44%
SH 38+12 East	14	1	3	moderate	0.073	0.070	0.070	4.11%	5.4	2973	6.4	13700	21	14	67%
SH 38+12 West	14	1	5	moderate	0.073	0.070	0.071	4.11%	5.4	2973	6.4	13700	21	14	67%
SH 44+50	16	3	3	none	0.057	0.049	0.055	14.04%	5.4	4212	7.5	24000	19	14	74%
SH 90+38 East	16	5	4	moderate	0.057	0.056	0.055	1.75%	6.3	3914	7.2	21898	23	14	61%

* Indicates standing water (below peak of corrugation) present at time of inspection

10

Figures 1 and 2 are graphs of the actual perforation of the pipe at the time of the field inspections vs. the percentage of galvanized pipe life as predicted by the California Method. These graphs were generated by conducting a linear regression analysis on all of the data for all three field investigations made as part of this study (Alabama, Oregon, and Maine). For the purposes of calculating a "life multiplier," the linear regression was forced through the point (0,0). This resulted in very poor fit coefficients.

The above technique using regression analysis was developed by Potter in Report No. FHWA-FLP-91-006. That study concluded that Aluminized Type 2 at the Natchez Trace Parkway and Santiam Highway would last six times longer than the life predicted for galvanized corrugated steel pipe (CSP) by the California Method.

Figure 1 illustrates the "Actual Perforation" compared with the "Percent California Predicted Life" of the 21 Aluminized Type 2 culverts that were part of the field investigations. Notice that the majority of these pipes are performing better than the California Method would predict for a galvanized culvert in the same environment.

In figure 1 there are three "outliers." Two are points from locations 310.0 and 310.1 from the Natchez Trace Parkway. The previous study labeled these two particular culverts as outliers, due to the presence of a highly aggressive effluent found in the drainage feature. During this investigation, the pH and resistivities measured were considerably higher (more benign) than that measured in the previous study. Some have suggested that the local soil conditions changed over the years. The initial pH of the effluent may have been lowered by soils in a borrow pit within the drainage area. These initial low pH conditions may have dissipated over time.

During our follow-up investigation, we were able to locate soils with pH below 5. Standing water on those soils had a pH between 5.8 and 6.7. The most recent rain had been a few days before the visit. Later investigations by others (including the local Federal Lands representative) again discovered soils with extremely low pH values (around 2.0). These soils were on the side of the stream bed feeding 310.0, approximately 305 m (1000 ft) upstream of the culvert. At the time of our follow-up visit, we provided the local Federal Lands representative with the equipment necessary to measure water pH in the culverts on a weekly basis. Figure 3 shows these data along with water depths in the culvert for a 2-month period. The data seem to support a lower pH associated with low flow in pipe 309.4. The data do not suggest that severely low-pH (<6) water is seen at any of the three pipes.

Locations 310.0 and 310.1 were the only two culverts at the Natchez Trace Parkway that were experiencing any kind of flow conditions at the time of the inspection. A possible alternative explanation for the advanced deterioration is a synergistic effect of the potentially corrosive Alabama soils and incidence of abrasion. The wear pattern on these culverts is indicative of the type of deterioration that occurs as a result of abrasive wear. This wear occurs predominantly on the upstream tangents of the corrugations. Figure 4 shows the interior of 310.0. Figure 5 shows a close-up of the invert of this culvert. Notice in both photos that the exposed steel and apparent rusting is more significant on the upstream side of the corrugation. This is consistent with abrasive damage. Chemical attack would be expected to affect both sides equally, as shown in figure 6, which is a photo of an Aluminized Type 2 pipe (not in this study) exposed to an aggressive environment in the absence of abrasion.

The third data outlier is that from the location in Dexter, Maine. This culvert is experiencing the highest abrasion of any of the Aluminized Type 2 pipes in this investigation. A high-velocity flow passes through this culvert carrying a rock bed load. Figure 7 shows an overview of the Dexter culvert.

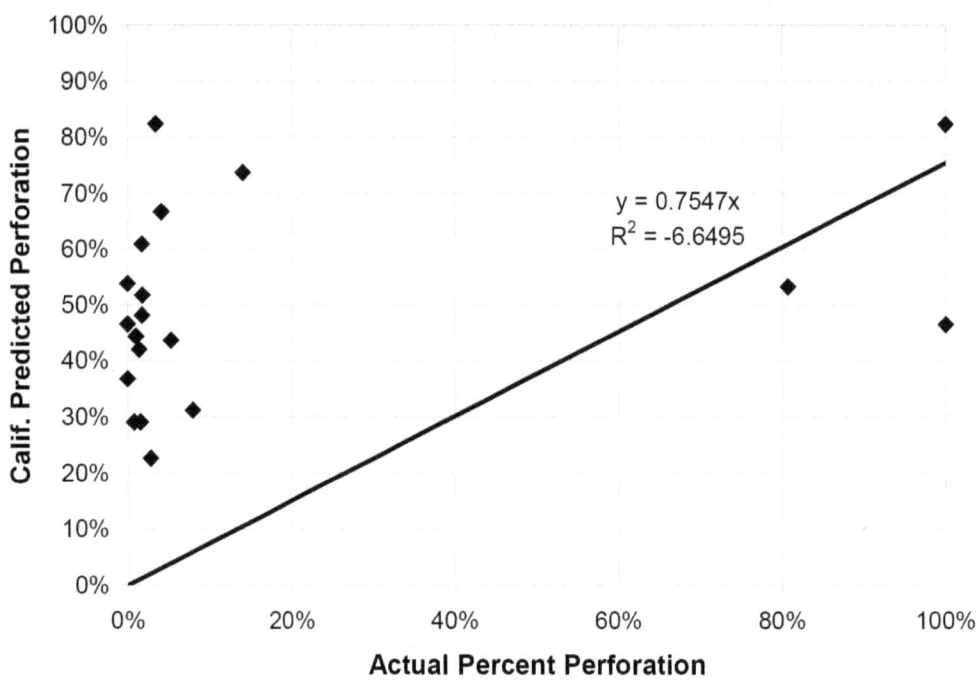

Figure 1. California Method curve prediction versus actual percent perforation, all Aluminized Type 2.

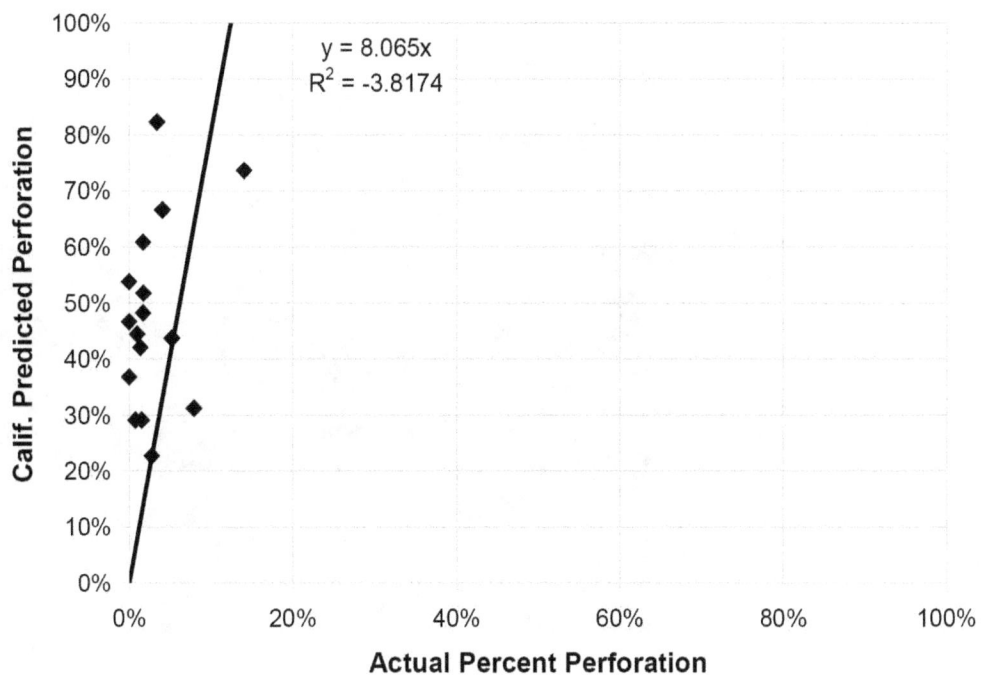

$$y = 8.065x$$
$$R^2 = -3.8174$$

Figure 2. California Method curve prediction versus actual percent perforation,
Aluminized Type 2 without outliers.

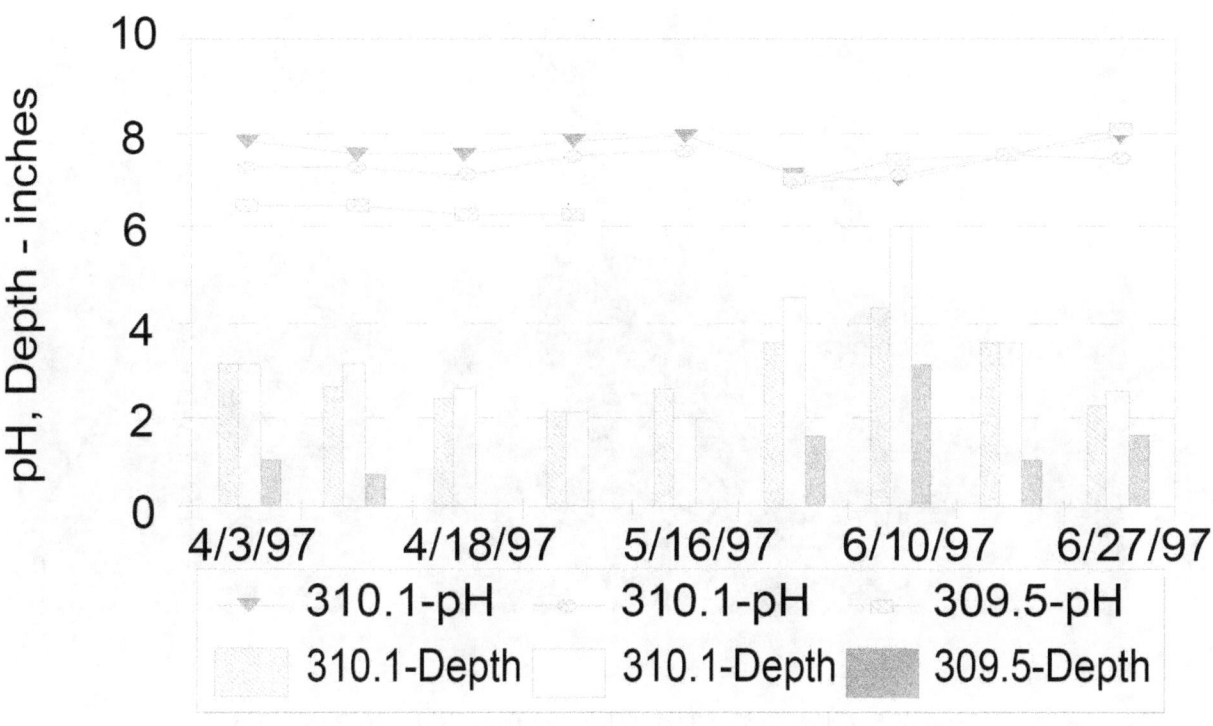

1 in = 2.54 cm

Figure 3. Water level and pH at the three
Natchez Trace Parkway pipes with perforation.

13

Figure 4. Culvert at 310.0 (flow is from bottom of photo to the top).

Figure 5. Close-up of 310.0 (flow is from left of photo to right).
Note: Wear and corrosion predominate on upstream side of corrugation.

Figure 6. Aluminized Type 2 culvert after 5.5 years
exposed to runoff with a pH of 2.5 to 3.5 (not from this study).

Figure 7. Culvert at Dexter, Maine, experiencing high-velocity flow.

Figure 2 shows the present data again, but without the outliers. Following a similar analysis, the data show that Aluminized Type 2 is eight times as durable as predicted for galvanized. The same type of analysis was used by the Florida DOT to conclude that Aluminized Type 2 might last 2.9 times longer than the life predicted for galvanized steel. The current analysis (without outliers) appears consistent with the previous work by Potter and the Florida DOT to the extent that it predicts a longer life for Aluminized Type 2 than would be predicted for galvanized steel.

When the outliers are included in the analysis, the advantage of Aluminized Type 2 is diminished considerably. Figure 1 shows the field data from this study. The difference between this study and the previous study is that this study has the five additional Aluminized Type 2 sites from Maine. When outliers are not included (figure 2), the advantage of Aluminized Type 2 is clear. When the outliers are included (figure 1), the advantage calculated in this study goes from 8 to 0.76; for the data of the previous study it goes from 6.1 to 0.44. The effect of data scatter on the resulting prediction is significant.

Figure 8 provides the same analysis as figure 2, yet uses the water-side chemistry to predict the California Method life. The correlation is no better. The ratio of actual life to the life predicted for galvanized drops from 8 to 3.5. This is expected because the chemistry of the water side appears less aggressive. A study of the plots in figures 1, 2, and 8 seems to show that, as noted previously, a poor fit is obtained by regression analysis and a better method is needed.

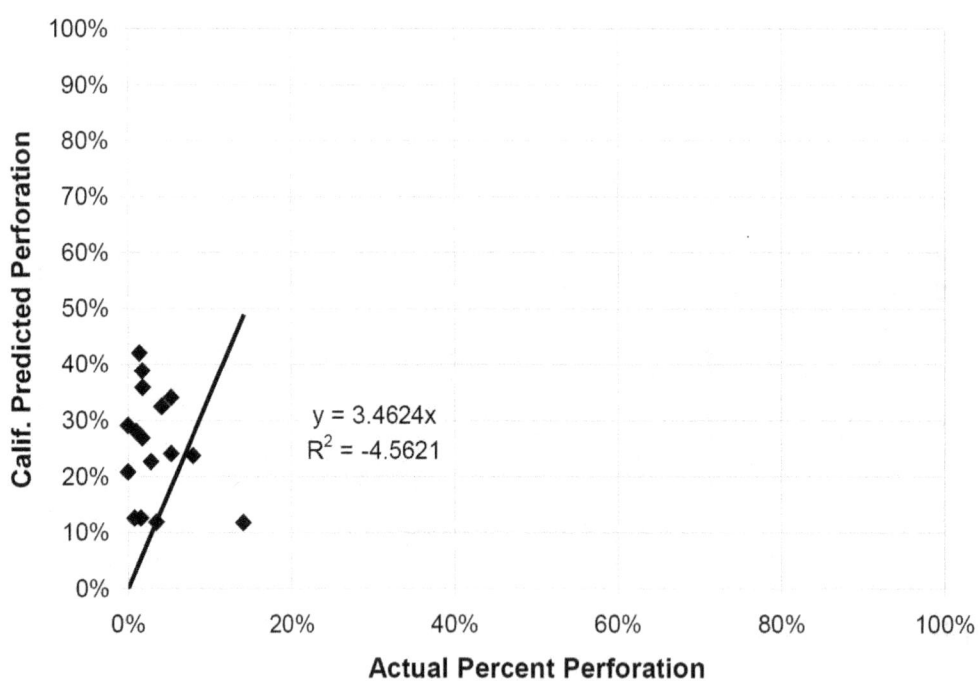

Figure 8. California Method predictions versus actual
percent perforation using water-side pH and resistivity.

The data scatter shown in figures 1, 2, and 8 suggest that more than one failure mode exists. It also suggests that failure cannot be predicted by pH and resistivity alone. It is also logical to expect that the perforation of the coating (roughly 5 percent of the pipe thickness) would occur at a different rate than perforation of the steel (remaining thickness). Given that the data obtained are maximum pit depths in each culvert, the pit-depth distribution may be amenable to other types of statistical analysis. For maximum values, an appropriate process might be an extreme value analysis.

In 1922, E.J. Gumbel, at Columbia University, first formally studied extreme values. His text treats this subject in depth (*Statistics of Extremes*, Columbia University Press, 1968). The present discussion gives a simplified explanation, limited to the application of extreme-value probability theory to the distribution of maximum corrosion pit depths. All engineers are familiar with the normal probability relationship and its bell-shaped curve. The curve, stated simplistically, is based on a totality of data — all of the scores on a test, average annual rainfalls, or all pit depths. To consider the probability of extreme values, one takes the totality of data and divides it into statistical groups — the highest scores of each group of students, the maximum daily rainfall of each year, or the maximum pit depth of each unit length of pipe. These new data, the extreme values, plotted on the same basis as the normal probability data, give a curve that is no longer symmetrical, but a bell that is skewed. The equation of this line will be:

$$Y = u + Ax \tag{1}$$

where,

Y = maximum value of pit depth
u = mode of pit distribution (y-axis intercept)
A = constant (slope of line)
x = (-ln(-ln(cumulative probability)))
Cumulative Probability = (sample rank)/(sample size + 1)

The equation states mathematically that, in a given system, the greater the percentage of the system examined, the greater will be the maximum pit depths. Pit depths systematically measured on less than the whole system can be used to predict the maximum pit depth that exists on the whole system. The equation also allows the calculation of probable number of pits in the system that will exceed any specified depth.

Application of the equation to practical situations requires a statistically valid collection of data. A practical and consistent sample size must be selected, and a sufficient number of samples must be taken so that the desired reliability of results may be attained. The deepest pit in each sample area must then be measured. These maximum pit depths form the database; regression analysis can then be used to determine the mode "u" and the constant "A" for equation (1).

Assume, for instance, that the total surface area inspected is divided into 100 equal sub-areas, which become statistical units. The maximum pit depth is measured on each area. Regression analysis is then used to determine the parameters of equation (1). Now assume that the total surface area of the structure of interest is 1,000 times the area of the surface inspected. The total structure surface area would therefore comprise 100,000 statistical units. The maximum pit depth most probably existing on the entire structure area would be the piece of information of first importance. The value of this maximum pit depth can be calculated by solving the equation using an assumed cumulative probability of 0.99999, corresponding to a return period of 100,000. If this value exceeds wall thickness, the number of pits exceeding the thickness would then be of major interest. This number could be determined from equation (1) by substituting the thickness value for X and solving for cumulative probability.

In the current study, we attempted to obtain the maximum pit depth from the invert in each culvert. Figure 9 provides a plot of the data from all of the Aluminized Type 2 sites in this study. Pit depths are the percent perforation times the theoretical thickness for the gauge of culvert.

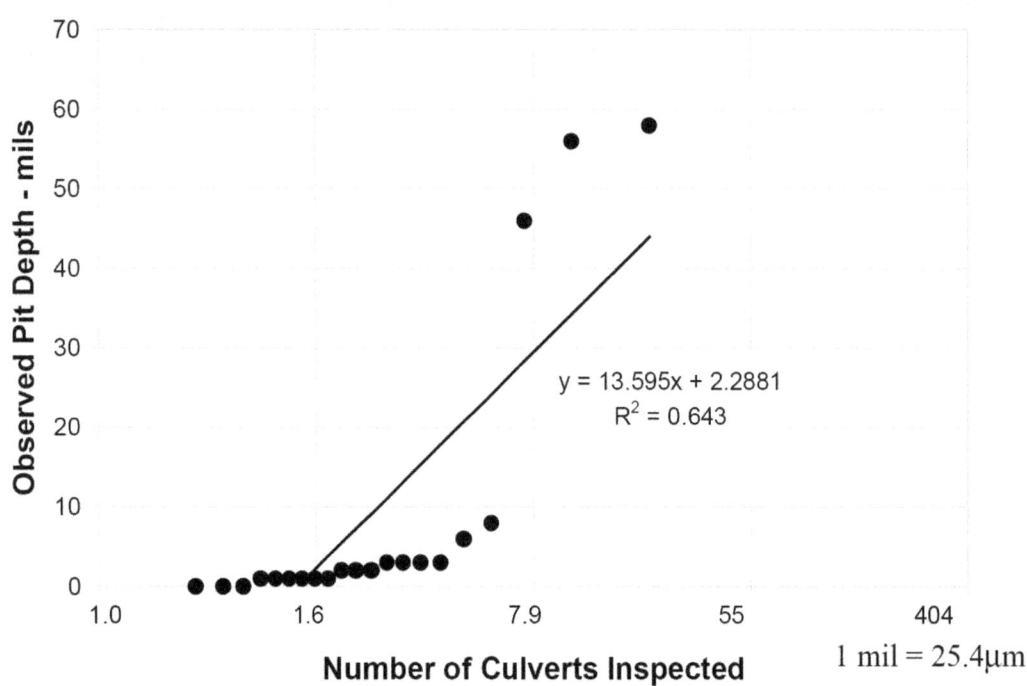

Figure 9. Extreme-value distribution for all Aluminized Type 2 data in table 1.

The data suggest at least two modes: a low and high pitting rate. Figures 10 and 11 show these distributions separately. The split between high and low pitting appears to be at about 127μm (5 mils). The samples in this study are between 10 and 17 years old.

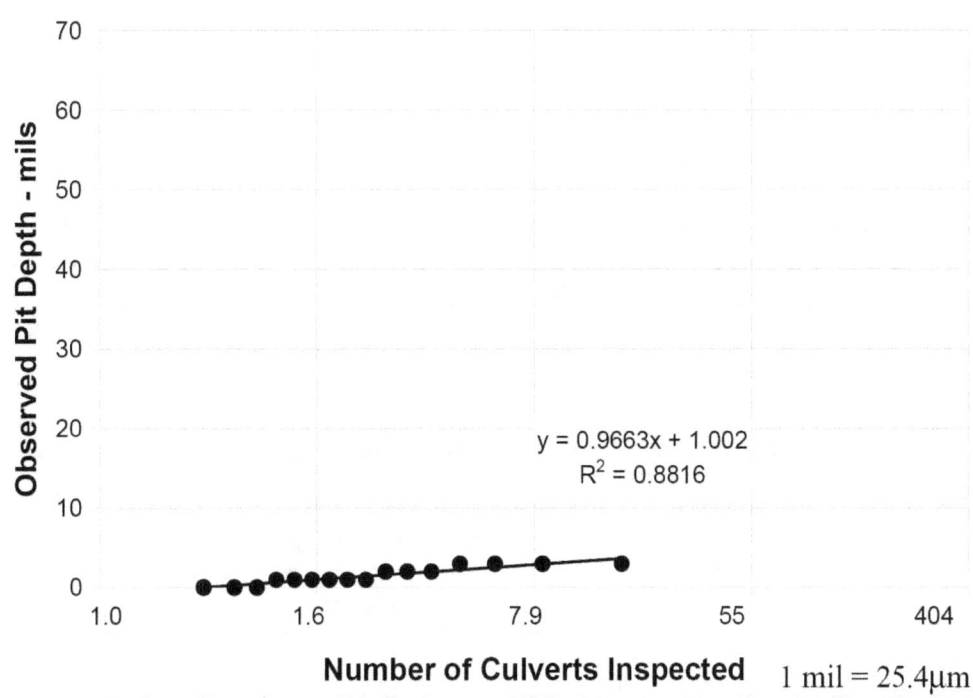

Figure 10. Extreme-value distribution for Aluminized Type 2 data in table 1 exhibiting a low pitting mode.

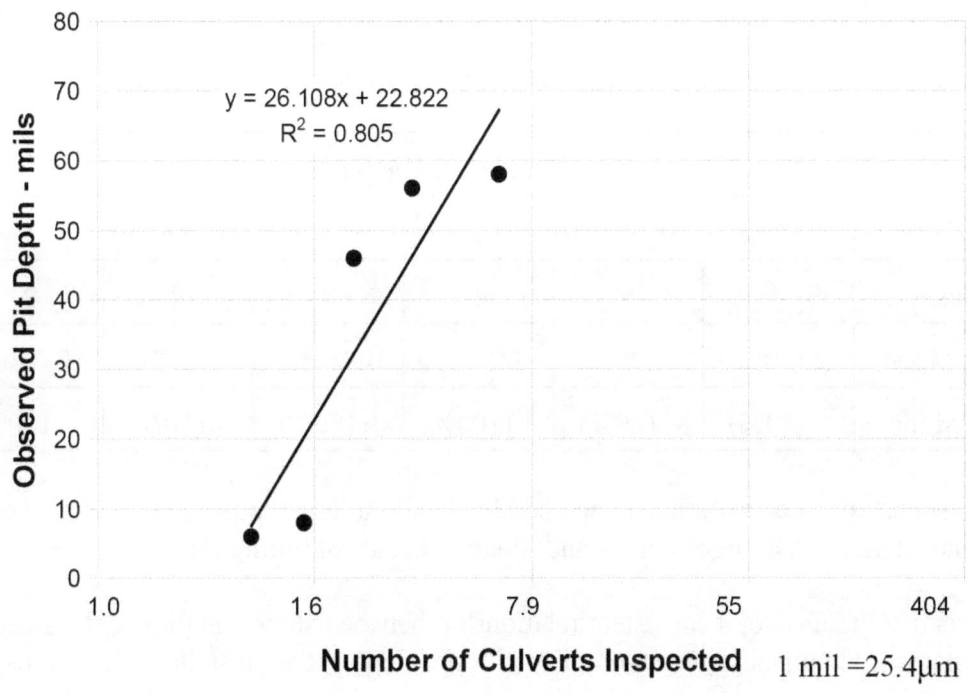

Figure 11. Extreme-value distribution for Aluminized Type 2 data in table 1 exhibiting a high pitting mode.

Each plot has a good correlation coefficient, suggesting that the data fit the model. It is also interesting that the split between low and high pitting mode places more than the three original outliers into a high-pitting rate group. There appear to be six sites (all those with more than 7.5 percent perforation in table 1) experiencing a high pitting rate. Conversely, there are 18 sites with a very low pitting mode.

Tables 2 through 4 examine the relationship between the flow conditions and the pitting rates. Each table shows a particular culvert characteristic across the top. The number of pipes exhibiting each level of the characteristic that fall into the high and low pitting mode is then indicated.

Table 2. Bed load versus pitting mode.

	Bed Load Characterization			
	None	Minor	Moderate	Heavy
Total Samples	7	7	6	1
High Pitting Mode	1 (14%)	1 (14%)	2 (33%)	1 (100%)
Low Pitting Mode	6 (86%)	6 (86%)	4 (67%)	0 (0%)

Table 3. Slope versus pitting mode.

	Culvert Slope (Degrees)					
	0	1	2	3	4	5
Total Samples	4	5	1	7	2	2
High Pitting Mode	2 (50%)	1 (20%)	0 (0%)	1 (14%)	1 (50%)	0 (0%)
Low Pitting Mode	2 (50%)	4 (80%)	1 (100%)	6 (86%)	1 (50%)	2 (100%)

Table 2 suggests a positive correlation between bed load and increased pitting rates. However, the bed load characterization alone is not a stand-alone indicator of pitting rate.

Table 3 suggests that there is not a consistent relationship between slope and pitting. This is not surprising, as many of the pipes are cross-drains with little slope yet are installed adjacent to hills, thus receiving significant abrasion.

Table 4 suggests that there may be a weak correlation with velocity (as observed during our inspection) but not as strong as the relationship with bed load.

Table 4. Velocity versus pitting mode.

	Velocity (ft/s)			
	0	1-3	4-7	>7
Total Sample	4	10	6	1
High Pitting Mode	0 (0%)	3 (30%)	1 (17%)	1 (100%)
Low Pitting Mode	4 (100%)	7 (70%)	5 (83%)	0 (0%)

1 ft/s = 0.305 m/s

Tables 5 through 8 provide a similar analysis between the chemistry data and pitting mode. The chemistry ranges were divided to put about the same number of samples in each group.

For the chemistry data, the only positive trends appear to be increased likelihood of a higher pitting rate at lower water-side resistivities. The other parameters do not appear to fit the pitting mode.

Table 5. Soil pH versus pitting mode.

	Soil Ph		
	5.3 - 5.7	5.8 - 6.7	6.8 - 7.6
Total Samples	9	6	6
High Pitting Mode	1 (11%)	4 (67%)	0 (0%)
Low Pitting Mode	8 (89%)	2 (33%)	6 (100%)

Table 6. Soil resistivity versus pitting mode.

	Soil Resistivity - $\Omega \cdot cm$		
	<3,000	3,000 - 4,650	>4,650
Total Samples	6	7	7
High Pitting Mode	1 (17%)	2 (29%)	2 (29%)
Low Pitting Mode	5 (83%)	5 (71%)	5 (71%)

Table 7. Water pH versus pitting mode.

	Water pH		
	6.4 - 6.9	7.0 – 7.2	7.3 - 7.9
Total Samples	8	9	5
High Pitting Mode	1 (13%)	2 (29%)	2 (40%)
Low Pitting Mode	7 (87%)	5 (71%)	3 (60%)

Table 8. Water resistivity versus pitting mode.

	Water Resistivity - $\Omega{\cdot}$cm		
	<9,150	9,150 - 19,000	>19,000
Total Samples	7	6	7
High Pitting Mode	2 (29%)	1 (17%)	2 (29%)
Low Pitting Mode	5 (71%)	5 (83%)	5 (71%)

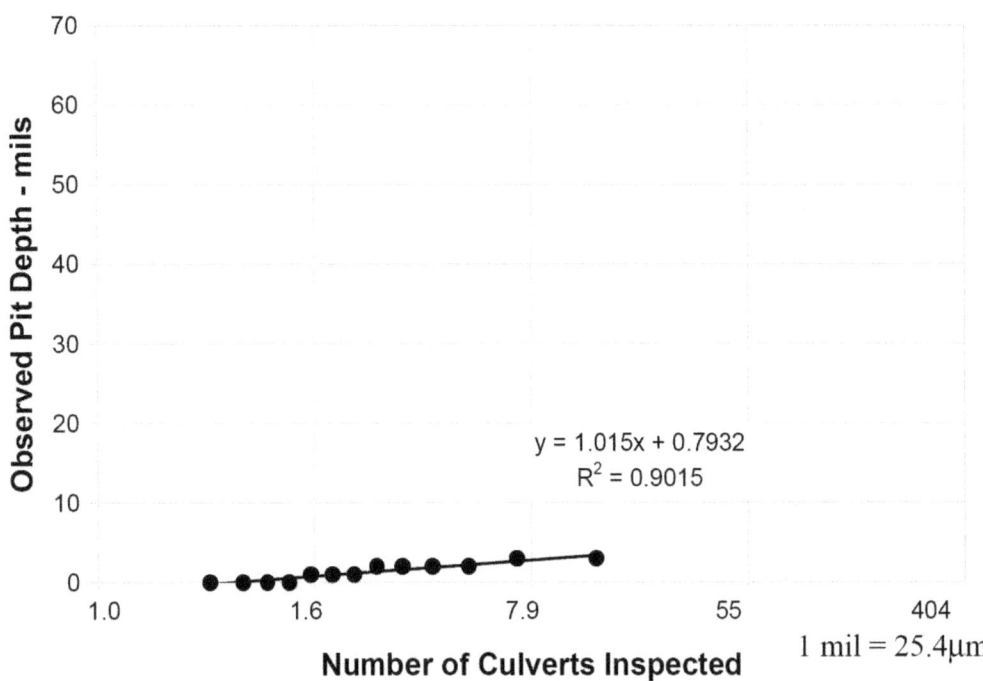

$y = 1.015x + 0.7932$
$R^2 = 0.9015$

1 mil = 25.4μm

Figure 12. Extreme-value distribution with use of
Potter's original data, low pitting mode.

22

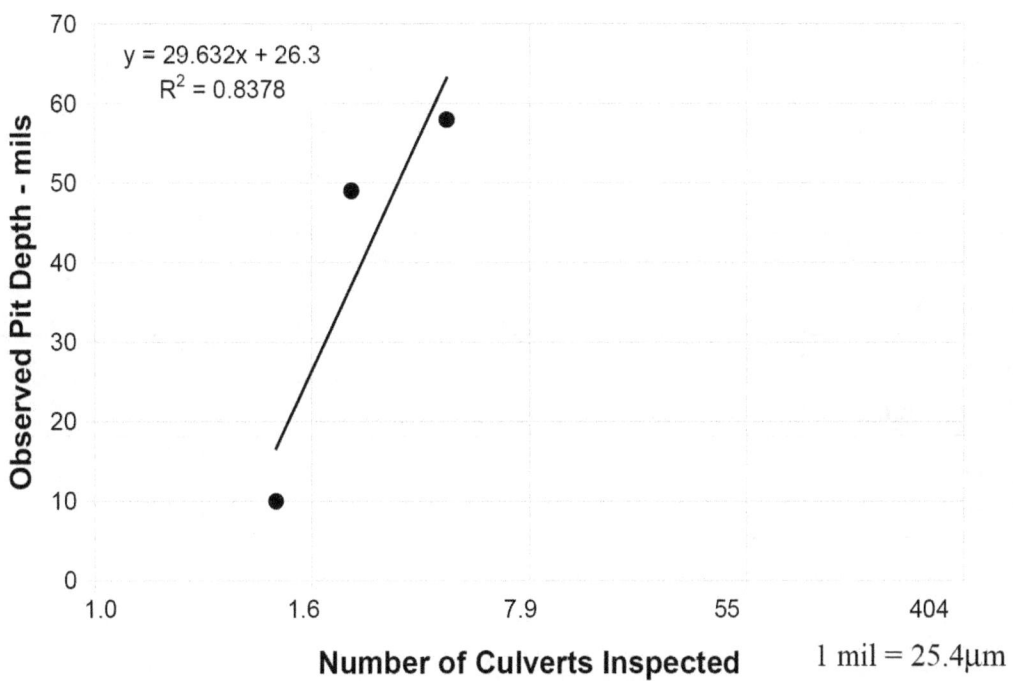

Figure 13. Extreme-value distribution with use of Potter's original data, high pitting mode.

Figures 12 and 13 show the culvert pitting modes described by Potter's original data, using the extreme-value analysis described above.

Again the data appear to follow an extreme-value distribution quite well. The data for each set may be compared through their mode and slope calculated by the best fit of the data. Table 9 provides this analysis.

Comparison of these data over time is interesting. In the low pitting mode, the mode of the data essentially doubles as the exposure period doubles. Figure 14 presents this relationship.

Table 9. Summary of extreme-value distributions.

Sample	Nominal Age – Year	Distribution Mode - "u" in Eq. 1	Distribution Slope - "A" in Eq. 1
Original Data Low Pitting Mode	7	0.79 mils	1.02
Original Data High Pitting Mode	7	26.3 mils	29.6
Current Data Low Pitting Mode	14	1.00 mils	0.97
Current Data High Pitting Mode	14	22.8 mils	26.1

1 mil = 25.4μm

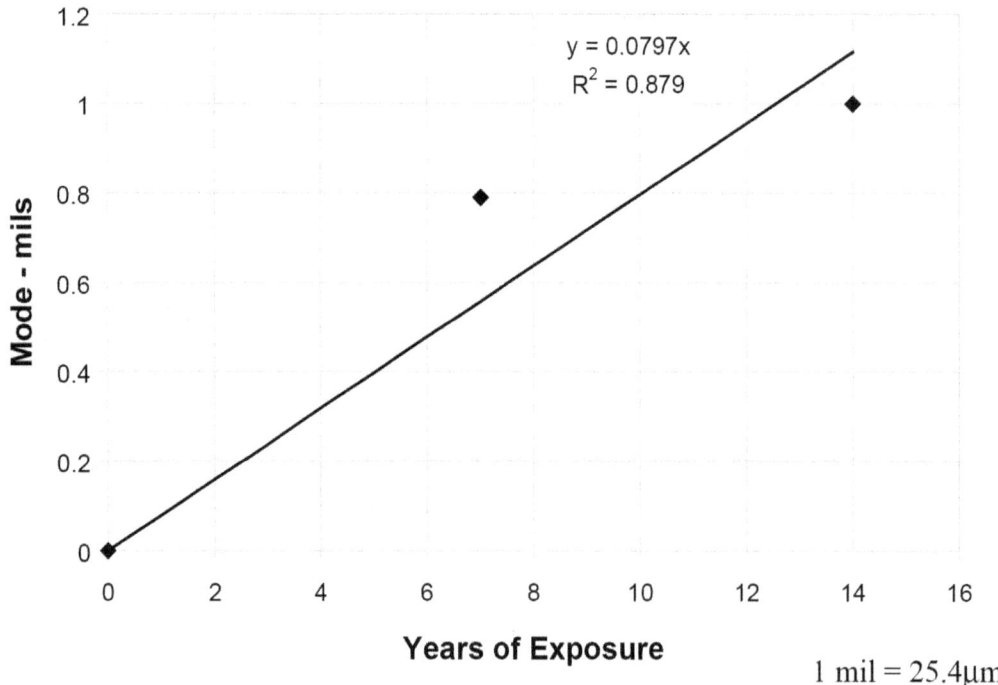

1 mil = 25.4μm

Figure 14. Mode of pit depth versus time.

For a coating thickness of 48.3 μm (1.9 mils), the analyses suggest that it would take more than 20 years for the most frequent, maximum pit to exceed the thickness of the aluminum coating. This is for the set of Aluminized Type 2 culverts examined in the current study that exhibit a low pitting tendency. Note that this is the time for the data mode to exceed 48.3 μm (1.9 mils); the first pit through the aluminum occurred in less than 7 years in the low pitting mode.

The data for the high pitting mode show little change over time. This is not surprising as the data are dominated by penetrations or near-penetrations. Penetrations cannot get deeper with time.

In summary, the Aluminized Type 2 CSP studied appears to be exhibiting at least two types of pitting. This results in two significantly different pitting rates. At the lower rate, the penetration rate for the sites investigated suggests that at least 20 years might be needed before complete penetration of the aluminized coating. At this point in time, additional life is provided by the remaining steel and any electrochemical protection provided by the aluminum. This life is very difficult to assess, but because we are concerning ourselves with an environment with a low pitting tendency, it would be possible to get at least another 40 years of life.

The higher pitting rate results in a more rapid perforation of the culvert. It appears slightly related to higher bed loads and lower solution resistivity.

Uncovered in the literature review were three other studies with data on galvanized steel culverts that were amenable to an analysis similar to that Report No. FHWA-FLP-91-006. These studies were conducted by the Florida DOT,[1] the Idaho Department of Highways,[2] and the Southeastern Corrugated Steel Pipe Association.[3] In total, data from 240 galvanized steel culverts from Florida, Idaho, and Georgia were collected into a database. Figure 15 shows a plot of percent thickness perforation versus percent of California Method predicted life (similar to those presented for Aluminized Type 2). All data were included in a regression analysis performed on these data.

From the analysis, it appears that this set of galvanized culverts is showing an actual service life increased by a factor of 1.9 versus the percent of California Method predicted life expended. This exercise shows that this analysis technique does not accurately correlate galvanized CSP condition (i.e., percent perforation) to pipe age as a percentage of California Method projection. The California Method is a guideline for predicting culvert durability developed and intended for use in California. Each State may make its own model based on the local parameters of concern.

It is tempting to compare the galvanized CSP pitting data with that of the Aluminized Type 2; however, the data were not obtained in the same environment so this is of little utility. The only points of environmental comparison between the Aluminized Type 2 and the galvanized CSP are the soil and water pH and resistivity measurements, which do not fully explain the corrosion behavior of CSP. Thus, data from both materials in comparable service environments should be compared rather than data for one material being compared with "other" predictions. There is a possibility that galvanized would perform as well as Aluminized Type 2 in these locations — showing higher durability than the California Method predicts. However, there was some evidence in Oregon and Maine showing direct advantages of Aluminized Type 2 over galvanized in a similar environment. At Garland, Maine, the galvanized section of the tandem installation showed more deterioration than the Aluminized Type 2. Also, some of the galvanized end sections on the Natchez Trace Parkway showed more deterioration than the aluminized culverts. Because of the limited amount of data available for the galvanized material the magnitude of this advantage was not quantified in terms of expected service life.

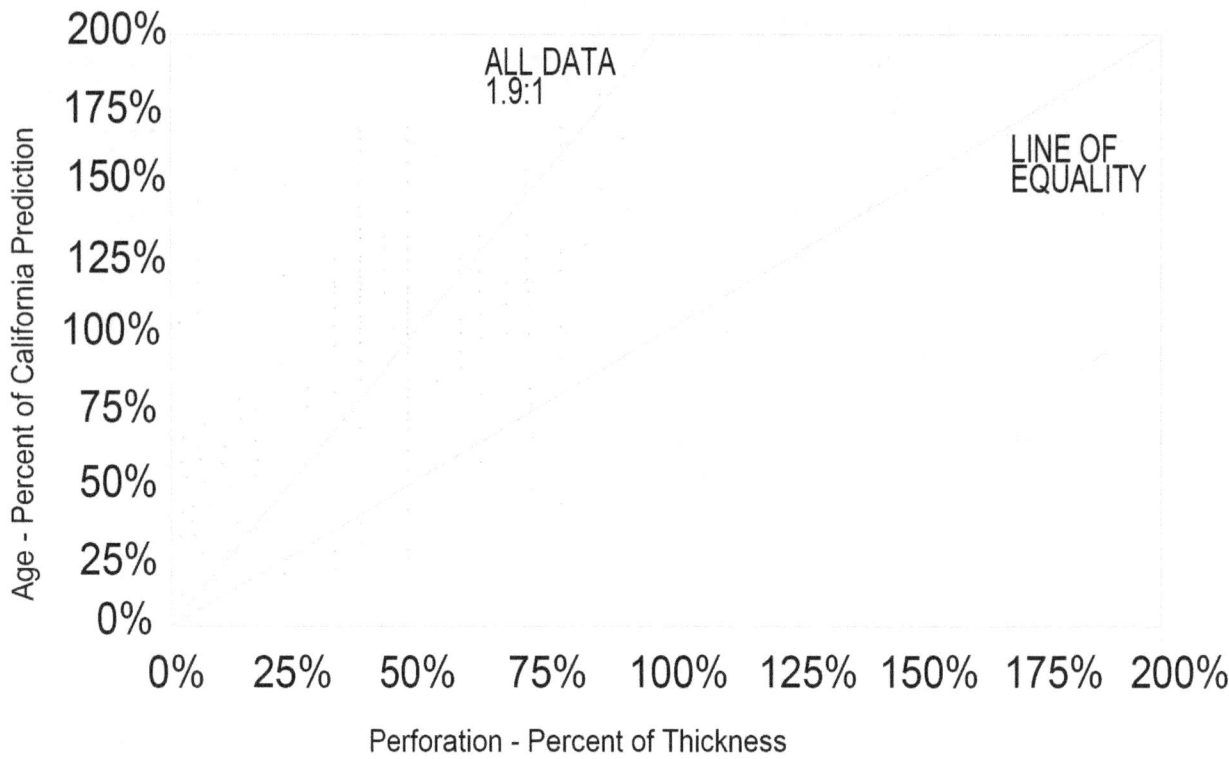

Figure 15. Reduction of galvanized data from three published studies.

The following discussion presents observations from the field studies conducted as part of this study.

Natchez Trace Parkway, Alabama

The Natchez Trace Parkway is a highway extending from the southwest corner of Mississippi through the northwest corner of Alabama. The sites investigated are in a 6.4-km-long (4-mi-long) area between mile markers 313 and 309 in Alabama near the Mississippi border. Some of these sites were difficult to find because of missing and inconsistent mile marker posts. This section of Alabama is hilly, but generally the culverts had flat to gradual grades at the entrances. Only a few culverts appeared to experience any significant abrasive flow. The existence of high mud stains on the majority of the pipes at this Natchez Trace Parkway location suggests that most of the sites seem to experience intermittent non-abrasive flood conditions. All of the sites that were showing obvious visual signs of deterioration (including 310.0, 310.1, 310.6, and 311.9) had indications that they may experience periodic abrasive flow.

At the time of the field studies on the Natchez Trace Parkway there had been two strong, quickly dissipating rain showers: one the week before the investigation was conducted and one that occurred the day of our arrival. The weather was hot [32.2°C (90°F)] and humid, which is typical for early summer in the South. Thus, the sites were relatively dry, allowing easy examination of the pipe condition. Two of the sites (310.0 and 310.1) appeared to be subject to abrasive flow at the time of our inspection. These sites were labeled by the previous study as outliers because of the extensive deterioration postulated to have been caused by the occasional presence of a corrosive effluent flowing through the pipe. During this investigation, evidence of a corrosive effluent was not found, although the site at 310.0 had a low soil pH. These locations had a consistent water flow with gravel present, indicating possible abrasion. The majority of the deterioration at these sites was on the upstream tangents, which are the areas that would be most affected by an abrasive flow. An aggressive effluent alone would affect the entire wetted area of the pipe.

The results of the Natchez Trace Parkway field studies show that the majority of pipes installed at this location are experiencing a mildly corrosive environment with soils having mainly neutral pH and low resistivities. The combination of low resistivities and the aggressive flows at 310.0 and 310.1 could possibly be the cause of the accelerated deterioration at these sites. Being within 4.8 km (3 mi.) of each other, the environmental characteristics of each site were similar. The pH of the soil and water averaged 6.8 and 7.3, respectively. The average resistivity of the soil as measured with the soil box was 3248 ohm-cm, and the average resistivity of the water was 8575 ohm-cm.

Most of the Natchez Trace Parkway installations are 16-gauge Aluminized Type 2 material. The California Method predicts that a galvanized steel pipe of 16-gauge construction will last 28.2 years until its first perforation in this environment. These culverts on the Natchez Trace Parkway are now 14 years old. Of the 11 culverts considered in the Alabama portion of this study, 2 out of 8 of the Aluminized Type 2 are perforated 100 percent across small areas of the invert. A bituminous-coated pipe at 310.6 is also in poor condition, with its bituminous coating missing in sections and with exposed, pitted steel. Detailed comments on each pipe follow.

312.4 West Pipe

This aluminum culvert is one of three culverts located on the access road to the Bear Creek recreation area at mile 312.4 and is the western of twin pipes that run north/south (parallel to the Natchez Trace Parkway). The pipe is of riveted construction with metallic end sections, a 106.7-cm (42-in) diameter, a length of 29 m (96 ft), and a corrugation size of 6.8 cm by 1.3 cm (2-2/3 in by 0.5 in). The weight and appearance of the coupons removed suggest that the pipe is made of aluminum alloy. The mud stains and high-water lines on the pipe led us to believe this pipe experiences some flood conditions. At the time of the study, this pipe had water above the 7:30/4:30 positions, making it difficult to observe the condition of the invert. The water extended from a stagnant pool at one end of the pipe to a stagnant pool at the other end. The installation appeared to be level, as observed by the same depth of water at both ends. This flatness and the lack of any defined stream bed suggest that this installation experiences very little, if any, flow.

Physical examination through the murky water found that the invert of this pipe felt smooth, which suggests that there is very little corrosion on the invert. The rest of the pipe is in good condition. Mud had stained the entire inner circumference — evidence that flooding must be experienced at other times of the year. It appeared as if the water at this site had been stagnant for some time. The pipe serves as a cross-drain between the sides of the access road.

312.4 East Pipe

This Aluminized Type 2 pipe is the eastern one of twin pipes crossing the access road to the Bear Creek recreation area at mile 312.4. The pipe is of welded seam construction with helical corrugations, metallic end sections, and physical dimensions the same as the west-side pipe. It has the same orientation, flow, and environmental factors as does the 312.4 west pipe. This is also a level installation in approximately the same condition as its neighbor.

312.4 Single Pipe

This Aluminized Type 2 pipe is located further down the access road. The pipe has a helical, welded seam with metallic end sections, a diameter of 45.7 cm (18 in), a length of 13.4 m (44 ft), and a corrugation size of 6.8 cm by 1.3 cm (2-2/3 in by 0.5 in). There was little water at this location — only a small amount between the pipe corrugations. The pipe is stained completely around its inner circumference with mud, suggesting flood conditions at other times of the year. After cleaning away the mud, the coating felt smooth to the touch and appeared to be in excellent condition. This is indicative of exposure in a fairly benign environment. This is a level installation whose surroundings indicate that it serves to equalize water on either side of the recreation access road. The pipe is probably not exposed to an aggressive flow. Again, as with other locations on the Natchez Trace Parkway, there is no clear definition of a stream bed and it is possible that this pipe remains dry for long periods.

311.9 North Pipe

This Galvalume pipe is the north side of a double run with concrete headwall and wingwalls. It has helical, lock-seam construction, a diameter of 182.9 cm (72 in), a length of 36 m (118 ft), and a corrugation size of 7.6 cm by 2.5 cm (3 in by 1 in). There was a 5.1-cm (2-in) collection of gravel/stone on the invert, suggesting that this pipe may experience periodic abrasive flow. The downstream portion of this pipe was covered across 0.6 to 0.9 m (2 to 3 ft) of the invert with small stones. The gradual slope of the pipe and flatness of the grade on the upstream end indicates that the velocity during flow times may not be very high. The pipe had minor invert corrosion over 45.7 cm (18 in) of its circumference and was rough to the touch. A high water line (3 o'clock) indicated that this pipe is half full at times. There was some evidence of a stream bed at wetter times of the year, but it was not clearly defined. During our follow-up investigation, it was evident that this pipe receives more abrasive exposure than the 311.9 south pipe, though it could only be classified as "minor" abrasion.

311.9 South Pipe

This Aluminized Type 2 pipe is the south pipe of a twin culvert with concrete headwalls and the same physical attributes as the north pipe. The absence of stones in this pipe suggests that the slope may be different or that the approach angle of the water during wetter times causes more flow in the northern pipe. Silt and mud were deposited between the corrugations. After cleaning the mud away there appeared to be very little corrosion on the invert and the coating was smooth to the touch. Again, similar to the north pipe, there was a high-water line (3 o'clock/9 o'clock position), suggesting that it experiences some intermittent flood conditions. During our follow-up investigation, it was evident that this pipe receives less abrasive exposure than the 311.9 north pipe, though it was also classified as "minor" abrasion.

310.6 North Pipe

This Aluminized Type 2 pipe is the northern one of a double culvert with a concrete headwall and wingwalls. It has a diameter of 152.4 cm (60 in), a length of 28 m (92 ft), and a corrugation size of 7.6 cm by 2.5 cm (3 in by 1 in). There was no water at the time of our inspection, but the presence of rocks of up to 2.5 cm (1 in) in the invert suggests minor abrasive flow during some periods. Viewed from the road above, it is apparent that a stream flows through this site from the adjacent field during wetter conditions. There is clear evidence of a stream bed extending into the downstream field that is now overgrown. The relatively flat grades of both the field and the pipe suggest that this flow may be not very rapid. Unlike the other sites, there is no high-water line in this pipe. At the time of this investigation, the only water existing at this site was between the corrugations and pooled at one end. A black film, approximately 38.1 cm (15 in) wide, existed on the invert. When the film was scraped off with a penknife, the visible coating had minor corrosion and appeared in good condition.

310.6 South Pipe

This bituminous-coated galvanized pipe is the southern one of a double culvert sharing a concrete headwall and wingwalls. It has the same physical dimensions as the north-side pipe. The concrete headwall was cracked near the crown and was spalling significantly in the invert. Water conditions were the same as 310.6 north. Although the pipe was dry during this inspection, small stones of up to 2.5 cm (1 in) were found in the pipe, suggesting minor abrasive flow during some periods. Standing water was observed in the corrugations and at the ends.

The pipe is bituminous-coated — a hot dipping process that leaves excess coating in the lower corrugations (the invert during dipping). When this pipe was installed, the section of the pipe with excess coating was installed as the invert in all of the sections except one, the very eastern section (note that this was not a pipe with a paved invert). In this section, the bituminous coating was cracked, flaking, and absent from 63.5 cm (25 in) of the pipe circumference. A black film was covering this exposed area. The black film and flakes were scraped away, exposing pitted steel. The rest of the pipe appeared to be in good condition, though the bituminous coating was missing in small areas. This observation demonstrates the advantage of a thicker bituminous coating. The deterioration of both the headwall and the pipe itself were considerably greater on this pipe than the north pipe.

310.2

This Aluminized Type 2 pipe is of welded-seam construction with a concrete headwall and wingwalls. It has a diameter of 137 cm (54 in), a length of 24 m (80 ft), and helical corrugations of 7.6 cm by 2.5 cm (3 in by 1 in). Though not part of the original study, the pipe was added at the time of the field investigation. The coating type was confirmed with quantitative chemical analysis. Similar in appearance to others we inspected on the Natchez Trace Parkway, this pipe had mud stains and a flat slope. The mud stains (present up to 3 o'clock) indicate that in wetter conditions water floods this area and the pipe acts to equalize the water level between both sides of the roadway. There is no evidence of an abrasive flow of any kind at this site. The only water at this location at the time of the study was lying between the pipe corrugations.

Judging by the slope of the inlet and the pipe itself, it does not appear that the pipe ever experiences abrasive flow. Minor corrosion exists on the crests of the corrugations circumferentially for 30 to 38 cm (12 to 15 in). Dents visible from inside the pipe appear to have occurred during installation.

310.1

The Aluminized Type 2 pipe at this site is of welded-seam construction with a concrete headwall and wingwalls. It has a diameter of 183 cm (72 in), a length of 25.9 m (85 ft), and corrugation dimensions of 7.6 cm by 2.5 cm (3 in by 1 in). The pipe has a flattened shape, perhaps due to improper installation or shifting of soil loads over time. This site is experiencing steady flow from a small stream though the water is not very deep. This pipe is probably subjected to continuous flow, though in drier conditions the crests of the corrugations may be exposed to the air. There were no high mud stains, suggesting that this pipe does not get the backup flood conditions like most of the other culverts investigated on the Natchez Trace Parkway. Stones up to 5 cm (2 in) were found in the invert. It appears that a good rainfall might subject the pipe to a moderate amount of abrasive flow. The coating on this pipe is in excellent condition except for 45.7 cm (18 in) of the circumference (in the invert), where it is stained black and perforated on the crests of most of the corrugations. The perforations are between 20 and 25 cm (8 and 10 in) wide circumferentially and are located on the upstream tangents over the entire length. There are many nodules located on the corroding section of the invert (adjacent to the perforations). The highly corrosive effluent discussed in the previous study was not found in the multiple environmental measurements taken during inspection of this culvert.

310.0

This Aluminized Type 2 pipe is of welded-seam construction with concrete headwalls and wingwalls. It has a diameter of 183 cm (72 in), a length of 58 m (190 ft), and helical corrugations of 7.6 cm by 2.5 cm (3 in by 1 in) in dimension. The culvert carries a steady flow of water coming from a small stream. About 9 m (30 ft) before the entrance of this pipe, two small streams merge. The flow is a constant 0.6 to 1.2 m/s (2 to 4 ft/s). The flow rate and the deterioration level at this site seem greater than at pipe 310.1. The lack of mud stains indicates that this pipe does not experience the back-up type flood conditions that most of the other pipes on the Natchez Trace Parkway seem to experience. The water is not very deep and in drier conditions may expose the crests of the corrugation to the air. The coating on this pipe is in excellent condition except for a 0.9-m- (3-ft-) wide section of the invert where there is evidence of severe corrosion and/or erosion. All visible (not buried under deposited rock) corrugation crests are perforated (100 percent) on the upstream tangents. The perforations are about 25 to 30 cm (10 to 12 in) wide. Rocks fill the invert of the downstream end starting 30 m (100 ft) from that end. The farthest downstream part of this pipe is covered by approximately 0.6 m (2 ft) of sand and rock. The stream bed at either end is mostly rock, giving it the highest abrasive potential of any site on the Natchez Trace Parkway. The highly corrosive flow discussed in the previous study was not found in the multiple environmental measurements taken during this field investigation.

309.5

This pipe is of welded-seam construction with bituminous-coated metallic end sections. It has a diameter of 106.7 cm (42 in), a length of 35 m (115 ft), and a corrugation size of 6.8 cm by 1.3 cm (2-2/3 in by 0.5 in). The pipe material was questionable because its condition did not correspond with that in Report No. FHWA-FLP-91-006. Qualitative chemical analysis of the coating confirmed that it was, in fact, Aluminized Type 2. The next section discusses 309.4, which was most likely the pipe reported in FHWA-FLP-91-006. The water at this site was at the top of the corrugations and pooled at both ends of the pipe. Minor corrosion was observed on the crests of the corrugations for about 30 cm (12 in) in the invert.

309.4

Several sources have reported varying conditions of the aluminized pipe at 309.5, which is described above. The discrepancy arises from the fact that this culvert and the one at 309.5 are located within 0.16 km (0.1 mi.) of each other. To try to explain these discrepancies, we made detailed observations of the southern-most pipe during our follow-up visit and identified it as 309.4. The pipe at 309.4 was perforated while the pipe at 309.5 was only 5 percent pitted. Neither of these pipes was included in the 1991 report.

During our second visit, coating type on the pipes at locations 309.5 and 309.4 were determined using a chemical spot test. The test was conducted in general accordance with Report No. DOD-HDBK-249. Using this technique, aluminum can be differentiated from zinc. The test showed that both pipes have an aluminum coating, most likely Aluminized Type 2. Samples previously removed from the pipe at 309.5 were positively identified as Aluminized Type 2.

Figures 16 and 17 show the pipes at locations 309.5 and 309.4, respectively. It is interesting that the pipe at 309.5 appears to be fed from roadway runoff while the pipe at 309.4 appears to be primarily fed from the forest to the west of the roadway. It was reported that this area was used as a borrow pit during construction. As reported above, we found no evidence of low pH soil roughly 61 m (200 ft) up the stream bed. There was bed load in the culvert at 309.4, which was not evident in the culvert at 309.5. The bed-load difference may arise from differences in the abrasive potential of the water source. The pipe at 309.5 appears to receive roadway runoff that is collected in a concrete channel and fed to the culvert. The pipe at 309.4 appears to receive runoff from the field to the west of the road, allowing a better source of bed load, which causes abrasive wear of the coating from the pipe. Table 10 presents water data gathered from the pipe at 309.4.

Table 10. Water chemistry observed at location 309.4.

Conductivity	77µS/cm
PH	7.1 @ 11.1°C (52°F)
Chlorides	7.5 mg/L
Alkalinity	19 mg/L
Carbon Dioxide	2.2 mg/L
Calcium Hardness	15.4 mg/L
Total Hardness	31 to 55 mg/L

Figure 16. Inlet of the pipe at 309.5.

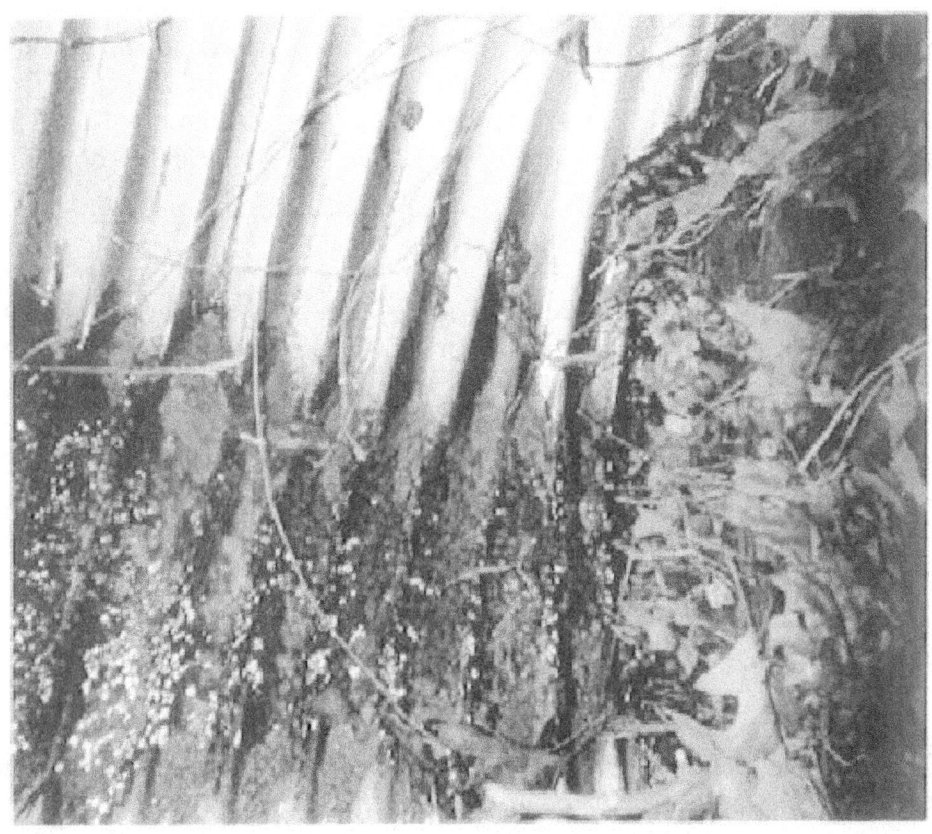

Figure 17. Inlet of the pipe at 309.4.

Santiam Highway, Oregon

The Santiam Highway is a section of Route 20. The culvert sites of interest are on a portion of this highway located 48 km (30 mi) from Albany between Foster Lake and Cascadia just outside the logging town of Foster. The sites are located within an 11-km (7-mi) portion of road that winds its way through a mountainous section of Oregon. All of the culverts have slopes between 1 and 5 degrees with flowing water. The purpose of these culverts is to channel the mountain runoff under the highway to the Santiam River. Most of these culverts have some amount of exposure to abrasive conditions because of heavily flowing water and rocky stream beds.

It was raining at the beginning of our inspection and continued to rain for the next 3 days. Since we conducted our inspection during this wetter time of the year in Oregon, this period was probably representative of the more severe flow conditions to which these culverts are exposed. A homeowner with property adjacent to one of the sites (119+20) informed us that the flow can increase considerably over what we saw. The heaviest flows observed at the Natchez Trace Parkway were mild compared with what was seen on the Santiam Highway.

The average soil resistivity measured at Santiam Highway using a soil box was 6565 ohm-cm, roughly twice that of the Natchez Trace Parkway. The average water resistivity measured was 26516 ohm-cm, nearly triple that of the Natchez Trace Parkway. The pH of the soils at the Santiam Highway sites were somewhat acidic, with an average of 5.9. The water pH measurements averaged a neutral pH of 7.2. Most of the culverts on the Santiam Highway are 16-

gage steel. The time to first perforation as predicted by the California Method for galvanized culverts subject to these average environmental conditions is 28.2 years. These culverts are now approximately 14 years old. Of the 13 culverts investigated, 11 have Aluminized Type 2 coating and all seem to be in excellent condition with no visual evidence of substantial corrosion.

13+00

This Aluminized Type 2 culvert is of welded-seam construction with a diameter of 61 cm (24 in), a length of 19.5 m (64 ft), and a corrugation of 6.8 cm by 1.3 cm (2-2/3 in by 0.5 in). Water flow is fairly consistent from a small stream that comes down a near vertical slope approximately 3 m (10 ft) from the inlet. The velocity of flow through this pipe at the time of this investigation was less than 0.6 m/s (2 ft/s). A few small stones [0.6 to 1.3 cm (0.25 to 0.5 in) diameter] were found in the valleys of the corrugations of the upstream portion. This site was given a minor abrasive rating because of the low velocity and volume flow. The invert appeared to have evidence of minor corrosion or staining, but the surface felt smooth to the touch.

18+20

At this site, the roadway appears to have been modified, creating the need to extend the Aluminized Type 2 culvert underneath the roadway with a larger diameter galvanized pipe (coating identified by manufacturer stamp on the side). The diameter of the upstream Aluminized Type 2 end section is 76 cm (30 in) and the diameter of the downstream (galvanized) end section is 91 cm (36 in). The length of the installation is approximately 53.6 m (176 ft). The corrugation size of the upstream end is 5 cm by 1.3 cm (2 in by 0.5 in) and the corrugation size of the downstream end is 6.8 cm by 1.3 cm (2-2/3 in by 0.5 in). The upstream end is of lock-seam construction with a bituminous-coated end section; the downstream section is also lock-seam, but with a square end and no end treatment. The bituminous-coated galvanized end section had coating missing from above and below the water line and the exposed metal surface showed severe corrosion in sections. At the time of the study, the flow covered about 0.3 m (1 ft) of the invert and had a velocity of approximately 0.6 m/s (2 ft/s). The moss growth and the stain pattern indicate that the flow may not get much greater. No abrasive material was found between the corrugations of either of the sections. When the moss growths and deposits were scraped away, the coating appeared to be slightly corroded on the upstream corrugations, but otherwise in excellent condition visually and smooth to the touch. The downstream (galvanized) section at the end of this pipe was on a greater grade because of a change in slope somewhere near the middle of the installation. The section at the downstream end of the pipe was severely corroded in the invert.

38+12 West Pipe

This Aluminized Type 2 culvert is of welded-seam construction with a diameter of 137 cm (54 in), a length of 21 m (69 ft), and a corrugation size of 6.8 cm by 1.3 cm (2-2/3 in by 0.5 in). The two culverts at this installation share a concrete headwall with wingwalls on the upstream end. The difference between the east and west culvert is due to the approach angle of the stream; the western pipe of the installation is receiving a greater volume at a higher velocity than the eastern one. The velocity of flow through this pipe was estimated to be approximately 1.5 m/s (5 ft/s).

34

The water was running over approximately 61 cm (24 in) of the invert. The coating on this pipe is in good condition although it was difficult to distinguish through the flowing water. There is dark staining on the downstream portion of the corrugations and moss growth on either end of the pipe. Unlike 38+12 East, the outlet of this pipe was not covered with stones. This may be because the volume and velocity of the water through this side is capable of flushing the rocks clear through the length of the pipe. This was supported by throwing handfuls of larger rocks into the pipe and observing them being carried through the length of the pipe.

38+12 East Pipe

The construction of this Aluminized Type 2 pipe was identical to the western pipe. The east end of this installation was covered with sand and rocks over about 0.9 m (3 ft) of the invert. Some of the rocks were as large as 20 cm (8 in). There was less flow through this pipe, with an estimated velocity of 0.9 m/s (3 ft/s). Although this installation was receiving less of the flow, the downstream end was filled with rocks to the 4 o'clock/8 o'clock position and up to 3 o'clock/9 o'clock with water. The amount and size of rocks collected in the downstream portion of this pipe are a clear indication that this pipe sees some abrasive flow. There was also evidence of wear on the crest of the corrugations.

44+50

This Aluminized Type 2 pipe is of lock-seam construction, has a diameter of 76 cm (30 in), a length of 23 m (76 ft), and a corrugation size of 6.8 cm by 1.3 cm (2-2/3 in by 0.5 in). The pipe has a bituminous-coated end section on the upstream end. The downstream end has no end treatment and is cantilevered over a steep incline that goes down to the river. The bituminous-coated end section has large sections of missing coating above and below the water line. The water line of the end section is corroded. This pipe has a moderate volume of flow that covers the upstream end to about 15 cm (6 in) over the invert. The velocity is higher and the volume is less on the downstream end because of a change in slope along the length of the pipe. The velocity upstream was approximately 0.6 m/s (2 ft/s) and the velocity downstream was approximately 0.9 m/s (3 ft/s). The coating is stained at the invert, but appears and feels to be in excellent condition. Abrasion does not seem to be a factor at this site.

90+38 West - Twin Pipe

This Aluminized Type 2 pipe is the western side of a double culvert. It was not listed in FHWA-FLP-91-006. It is of lock-seam construction, has a diameter of 91 cm (36 in), a length of 22 m (71 ft), and a corrugation size of 5 cm by 1.3 cm (2 in by 0.5 in). The upstream end treatment is a concrete headwall and wingwalls, and the downstream end is step-beveled. The inverts on both ends of this pipe were backed up with rocks. The upstream end was filled to the 4 o'clock/8 o'clock position with rocks and to 3 o'clock/9 o'clock with water.

The downstream end was somewhat less filled. Some rocks were as large as 25 cm (10 in). Although the water and rocks made it difficult to inspect the coating, it appeared to be in good condition and was smooth to the touch. The amount of water and rocks in the upstream end made it necessary to remove the coupons from farther downstream. The velocity of the flow at this site was estimated at 1.2 m/s (4 ft/s).

90+38 East - Twin Pipe

This Aluminized Type 2 pipe is the eastern side of a double culvert. This pipe is similar to its neighbor except that it has a welded seam and has a corrugation size of 6.8 cm by 1.3 cm (2-2/3 in by 0.5 in). Only the downstream end has collected rocks — some as large as 25 cm (10 in). The flow condition in this pipe is similar to the pipe at 90+38 West. The condition of the coating is also similar.

100+15

This Aluminized Type 2 pipe is of helical, lock-seam construction with a diameter of 76 cm (30 in), a length of 20 m (66 ft), and a corrugation size of 5 cm by 1.3 cm (2 in by 0.5 in). The upstream end has a bituminous-coated end section. The downstream end is square and cantilevered over the steep slope that leads down to the river. The bituminous-coated end section seems to be in good condition with its coating intact. An interesting observation at this site was the thick moss growing on the bituminous coating above the water line. This thick moss could be a reason why the bituminous coating is missing in large sections above the water line at site 44+50. There was bituminous coating missing from below the water line of this pipe. The flow velocity was estimated at 1.2 m/s (4 ft/s). There were no rocks found in the invert of the pipe. The flow was evaluated by tossing a handful of 5- to 7.6-cm (2- to 3-in) rocks into the culvert and observing them being carried through the length of the installation. Although the flow is capable of carrying small rocks, due to the layout of the site it did not appear as if many made their way through. This pipe appears and feels to be in excellent condition.

104+45 West Pipe

The location of this Aluminized Type 2 pipe is 1 km (0.6 mi.) past mile marker 39. Both of the structures located at this site are pipe arches. The pipe arch is of helical corrugation, welded-seam construction with a cross-sectional measurement of 180 cm by 119 cm (71 in by 47 in), a length of 18.6 m (61 ft), and a corrugation of 6.8 cm by 1.3 cm (2-2/3 in by 0.5 in). This pipe arch is step-beveled on both ends. A large volume of water flows through this pipe arch covering all of the lower part. The velocity was estimated at 1.5 m/s (5 ft/s). Of the Aluminized Type 2 culverts inspected in Oregon, this one has the fastest and heaviest flow. Removal of the coupons was considerably more difficult here because this pipe arch was made of heavy-gauge material, probably 10 gauge. Although there was much abrasive material at the site (ranging from sand to 25-cm [10-in] rocks), there was none found in the pipe arch. Apparently the flow at this location is capable of carrying stone through the length of the structure. This was confirmed by placing some large rocks in the invert and observing them flow through the structure. This side of the double installation was experiencing a heavier and faster flow than the other. There was dark algae-type growth/stain on the downstream side of the corrugations. There was a lot of growth on the flat portion of the arch in the exposed beveled ends. There was also evidence of what industry people refer to as "blushing." The blushing looks like minor corrosion of the crests of the corrugation over the entire circumference of the pipe arch. In spite of this blushing, the coating seemed to be in excellent condition.

104+45 East Pipe

This Aluminized Type 2 pipe is the eastern structure of this installation and has the same physical characteristics as its neighbor. The difference between the two is that, due to the approach angle, the velocity and amount of flow through the western culvert are greater than through this one. The flow through this pipe arch covered approximately 0.9 m (3 ft) of the invert. The flow on this side had considerably less velocity and volume than did the west side. There were no rocks found in the invert of this pipe arch.

113+25

This is a bituminous-coated culvert of welded-seam construction with a diameter of 107 cm (42 in), a length of 31 m (102 ft), and a corrugation size of 6.8 cm by 1.3 cm (2-2/3 in by 0.5 in). There is no end treatment on this pipe; upstream and downstream ends are both square. The first thing one notices about this site is that the bituminous coating is cracked and flaking from the exposed sections (near the ends and on the outside) of the pipe. The water at this site differed from the other sites in that it was brown and looked as if it had churned its way down the stream bed. The stream bed at this site is eroding substantially, sending rocks and sand through the pipe. When we moved some sticks and debris that had been blocking the upstream end of the pipe, the stream bed that was built up began to break up and dislodge sand and stone. In the time that we were doing our inspection, the stream shifted course and sent several stones down the length of the installation. It was apparent that this site experiences a considerable amount of abrasion. Rocks up to 10 cm (4 in) were easily carried through the pipe. The appearance of the stream and the surrounds seems to suggest that stones and debris are carried through this pipe on a regular basis. The flow at this site was heavy and covered the invert of the pipe to about the 4 o'clock/8 o'clock position. The velocity at this site was estimated at 1.8 m/s (6 ft/s). The bituminous coating was missing in patches between the 4 o'clock/8 o'clock positions. Corrosion was present on the crests of the corrugations.

119+20

This installation was the other bituminous-coated culvert inspected in Oregon. The pipe is of lock-seam construction with a diameter of 183 cm (72 in), a length of 27 m (88 ft), and a corrugation size of 12.7 cm by 2.5 cm (5 in by 1 in). There are no end treatments on the pipe and the downstream section is cantilevered. At the time of the investigation, this pipe was experiencing a high volume of flow. The velocity of this flow was estimated at 2.1 m/s (7 ft/s). A homeowner whose property was adjacent to this site informed us that the flow conditions become worse at other times of the year. Most of the coating was missing between the 3 o'clock and 9 o'clock positions. There was corrosion present on all of the upstream tangents. The flow at this site is comparable to the Dexter site in Maine.

123+76

This Aluminized Type 2 culvert is of lock-seam construction with a diameter of 91 cm (36 in), a length of 26 m (86 ft), and a corrugation size of 5 cm by 1.3 cm (2 in by 0.5 in). There are no end treatments on this pipe. At the time of the study, the flow at this site was fairly insignificant. The velocity at this installation was estimated to be 0.6 m/s (2 ft/s) and the flow was over approximately 20 cm (8 in) of the invert. It appeared as if abrasion at this site is not much of a factor. There was a small amount of moss-type growth at both ends of the pipe. The coating appeared and felt to be in excellent condition.

Maine

To expand the field studies of this investigation, several sites around the country were considered. Previously, Maine DOT established site installations as part of research it had been conducting. Those sites fit the criteria for extra sites defined within the contract. The current contract objectives required five Aluminized Type 2 sites to be added to expand the field studies. The sites in Maine are tandem installations of more than one type of material. As a result, it was possible to investigate galvanized zinc and Galvalume materials present in the same installation as the Aluminized Type 2. This type of installation provides for a convenient cross-comparison of materials subjected to the same environmental conditions. Most of these sites were part of the field investigation in the previous study, but comments were made on their conditions only and nothing was presented concerning environmental data. These sites are located near Interstate 95 between the cities of Portland and Bangor.

Prior to our visit, Maine had not had rain for 3 weeks. The evening before we began the field studies it began to rain and continued to rain through the next 2 days. Several inches of rain fell during this period, providing the opportunity to see the flow conditions to which these pipes are subjected. All of the sites were submerged to some degree under either flowing or standing water. At two of these sites (in the towns of Benton and Milford), an abundance of standing water made it impossible to collect the proper data and samples. As a result, these sites were not included in the field studies. Judging by the installation methods and the level of standing water at these sites it was clear that the purpose of these culverts is to equalize the water level on either side of the roadway. The sites at Garland and Orrington were added instead.

Most of the sites had relatively mild environmental conditions. Soil pH and resistivity measurements averaged 6.2 and 7823 ohm-cm, respectively. The average water pH and resistivity were 6.8 and 7965 ohm-cm, respectively. Using the worse case for pH and resistivity levels, the California Method predicts that a 16-gage galvanized pipe will last 30 years. The pipes at Orrington and Garland were 10 years old and the rest of the installations were 16 years old at the time of the inspections.

Dexter

The Aluminized Type 2 pipe located at this site is of lock-seam construction with a diameter of 213 cm (84 in), a length of 19.5 m (64 ft), and a corrugation size of 12.7 cm by 2.5 cm (5 in by 1 in) and the end sections are step-beveled. Of all of the sites in Maine, this pipe had the heaviest volume and the fastest flow. The flow in this pipe covered the downstream end to a depth of 45.7 cm (18 in). The velocity of this flow was estimated at approximately 1.2 to 1.8 m/s (4 to 6 ft/s).

Although the velocity does not warrant, this site was given an abrasive potential rating of level 3. This rating was given because stones up to 25 cm (10 in) were in the invert of the pipe. Obviously a higher velocity than was observed at the time of our visit would be necessary to move this size rock. The condition of this pipe is not good. There is moderate to severe corrosion taking place on 0.9 to 1.2 m (3 to 4 ft) of the circumference along the entire length. Another observation was the distorted shape of the pipe. This is most likely due to poor backfill procedure. The backfill used at this location contained very large rocks.

Garland

This site proved to be interesting because it was a tandem connection of pipes of three different materials. The upstream section was Galvalume, the middle section was galvanized zinc, and the downstream section was Aluminized Type 2 (there is no reason to believe that significant galvanic effects would occur in this arrangement). All sections have a diameter of 91 cm (36 in), a 15.8-m (52-ft) installation, and a corrugation size of 6.8 cm by 1.3 cm (2-2/3 in by 0.5 in). There is no end treatment. On the downstream end in the Aluminized Type 2 sections, approximately 15 to 20 cm (6 to 8 in) of rock, sediment, and sand had accumulated on the invert portion. The rocks are up to 7.6 to 10 cm (3 to 4 in) in size. The inlet slope at this installation is relatively flat and the velocity in the pipe is low (<0.6 m/s [2 ft/s]). It does not appear that the abrasive level could be very high. This installation was given an abrasive rating of level 1 to level 2. Again, as with other sites in Maine, the backfill consists of large rocks/boulders.

The variance in coating conditions was surprising given the moderate to mild flow conditions. It was difficult to inspect the condition of the coating because the pipe was only 0.9 m (3 ft) in diameter and filled with water. The Galvalume seemed to be in good condition, but had some areas with minor corrosion. The galvanized section had areas with moderate/severe corrosion. The pipe here was in visibly bad condition and rough to the touch. The Aluminized Type 2 appeared to be in very good condition. The coating was in visibly good condition and felt smooth. Another observation made was the condition of the re-rolled ends. The manufacturer of these pipes did not coat the re-rolled sections after the rerolling process. This caused considerable damage as evidenced by severe corrosion in the re-rolled/banded sections of these pipes.

The Aluminized Type 2 section of this pipe was withstanding local conditions better than either of the two other sections. The Galvalume section was not much worse than the Aluminized Type 2, with only minor evidence of deterioration. The galvanized pipe was in the worst condition of the three, having moderate corrosion, but no perforation.

New Gloucester

This Aluminized Type 2 pipe is of lock-seam construction with a diameter of 213 cm (84 in), a length of 18 m (60 ft), and a corrugation size of 12.7 cm by 2.5 cm (5 in by 1 in). The velocity of the flow at this site was approximately 0.6 to 1.2m/s (2 to 4ft/s) with a moderate volume covering about 0.9 m (3 ft) of the invert. This pipe was given an abrasive level of 1 because of the lack of velocity and abrasive potential. One end of the pipe had a headwall made of brick and some of the backfill consisted of old tires. However, much of the backfill material was large rock similar to the other Maine sites.

Orrington

This is another tandem installation with Galvalume downstream and Aluminized Type 2 upstream. Both sections of pipe are of lock-seam construction with a diameter of 91 cm (36 in), a length of 12 m (40 ft), and a corrugation size of 6.8 cm by 1.3 cm (2-2/3 in by 0.5). There was a very low velocity flow at this site, with little abrasive material present. Rocks were found deposited in the invert of the pipe near the middle and covered the invert to a depth of 7.6 cm (3 in). Because of the low velocity, this installation was given a level 1 abrasive rating.

Ripley

This Aluminized Type 2 pipe was of welded-seam construction with a diameter of 122 cm (48 in), a length of 27 m (90 ft), and a corrugation size of 6.8 cm by 1.3 cm (2-2/3 in by 0.5 in). The ends of the pipe were square with no conventional end treatment. Instead, a wire bed frame covered one end of the pipe. We learned from a local homeowner that beaver kept damming the pipe and to prevent this, the wire bed frame was placed at the upstream end. There is very little abrasive material present at this site and the velocity is approximately 0.6 m/s (2 ft/s). The coating on the pipe seems to be in excellent condition. The exception to this is the re-rolled ends of the pipe. As with other sites in Maine, the re-rolled ends were not recoated by the manufacturer. This resulted in premature deterioration of these re-rolled ends. The deterioration was not limited to the invert portion but was apparent around the entire circumference of the pipe.

LITERATURE REVIEW

Summary

A search was conducted to identify pertinent work of Federal and State agencies, industry, and universities. Many sources of research data were sought, including computer databases and recent publications. Some of the results reviewed include results published by the Federal Highway Administration (FHWA), U.S. Army Corps of Engineers, National Association of Civil Engineers (NACE), National Corrugated Steel Pipe Association (NCSPA), National Technical Information Service (NTIS), American Concrete Pipe Association (ACPA), and others. Phone interviews were conducted with various State agencies on their research and general practices with culverts. More than 140 papers were considered in the preliminary portion of the literature review; of these papers, roughly 60 were reviewed in detail. An annotated bibliography of these papers is included in appendix A. References will be made to selected literature review items throughout the text of this report.

One of the most common problems encountered while researching culvert durability issues is that there is very little standardization in the methodology used to test and evaluate culverts. Most research involves qualitative comparison of two or more products in "similar" installations or laboratory tests. Comparison testing is useful for an end-user evaluating alternative products for possible use. However, it is of limited use to other end-users if the specifics of the test installations are not adequately defined or controlled. Unfortunately, it is extremely difficult to adequately define test environments when no standard methodologies for such definition exist. These comparison tests typically provide visual inspection based on criteria set up by the investigator, but little environmental data. Visual inspections typically lack consistency when inspections are carried out by multiple inspectors with differing biases. It was our original intent to begin some sort of database to be included with this study, but lack of any common threads among many of the studies made it difficult. Abrasion, flow, and environmental criteria were not quantified in a consistent fashion. A major part of the current study is based on measurements of the deepest pits from removed coupon samples, allowing for a percent perforation comparison with existing life prediction criteria. Very few other studies had any analysis of pit depth.

Culvert durability prediction methods uncovered in the literature review include: AISI, NCSPA, California, Colorado, Florida, New York, and Utah methods. The California Method is most widely used and will serve as a baseline for comparison in this study. The AISI and Florida methods closely resemble the California Method. Charts and indices used for each method will be presented in the "Durability Prediction Methods" section of this report. These methods generally do not give significant consideration to flow conditions or abrasive levels.

This section is divided into four discussion areas: corrosion mechanisms and parameters, culvert materials and protective coatings, durability estimation procedures and examples, and software for estimation of durability.

Corrosion Mechanisms and Parameters

Most corrugated metal pipe (CMP) deterioration is due to a combination of corrosion and abrasion. The relationship between corrosion and abrasion is not well defined. Furthermore, it is a relationship that will vary for different materials. All metals form some type of oxide layer when they corrode (regardless of whether they are protective metallic coatings such as aluminum and zinc or base materials such as steel). Generally this oxide layer will have different properties than the base material. Typically, the oxide is more stable and its buildup will result in a decreasing corrosion rate. However, when the oxide is removed by an abrasive bed load, fresh underlying metal is exposed that will corrode. This cycle of corrosion and abrasion can thus result in higher net corrosion than in situations where the oxide layer builds up. Further complicating this corrosion and abrasion cycle is the fact that oxides typically have different levels of abrasion resistance than their base metal. Thus, the oxide buildup may help prevent metal loss by the increase in abrasion resistance it provides. In the case of steel corrosion, the corrosion product is less resistant to abrasion than the parent metal. This contributes to the severity of the corrosion and abrasion cycles.

The literature review contained discussions of the advantages of corrosion-inhibiting oxide layers forming on the surface of aluminum. It is not the intention of this study to deny the existence of these films or reject their usefulness. These films have been observed during this field investigation. However, the abrasion resistance of various oxides relative to the type of abrasion that occurs in the field has not been well documented. For example, it is certainly known that aluminum oxide is harder and more abrasion resistant than aluminum, but both aluminum and its oxides are susceptible to abrasive deterioration under heavy bed loads such as those observed in some culverts.

Typical corrugated steel pipe coatings are metallic coatings that are intended to provide sacrificial protection to the steel substrate (e.g., zinc, aluminum, and their alloys). However, one must remember that these coatings will self-corrode in addition to corroding by providing sacrificial protection. On pipe with relatively little exposed steel, the coating self-corrosion rate will often control the coating's service life more than the coating's sacrificial corrosion rate. Thus, selection of a coating that is a more efficient anode may not be prudent unless it also has a lower corrosion rate in the environment of interest. In this regard, corrosion rate data for the coating in the waters from the field environment would help to define the service life.

Where culverts fabricated with different materials or coatings are attached, the potential for galvanic corrosion exists. For example, there are many locations where galvanized and aluminized culverts (or culverts and end sections) are attached. Since zinc exhibits a more active electrochemical potential in most natural waters than aluminum, the zinc might be expected to corrode to protect the aluminum. However, the electrochemistry of this interaction depends on a number of factors, including electrochemical polarization of the metallic surfaces in question, oxide films formed on the surfaces, circuit resistance, current attenuation, and contact resistance between the two pipe sections. Each of these factors will create electrochemical inefficiencies that will reduce the galvanic corrosion.

Calculations based on data obtained in our laboratory suggest that maximum galvanic corrosion in an ideal electrochemical cell could reach 25 um (1 mil) per year of zinc. That would mean complete dissolution of the zinc within 2 to3 years. None of our field observations corroborated that this level of galvanic corrosion existed for a significant period of time.

Galvanic interaction would have been evidenced by a decreasing level of zinc deterioration observed further from the union between the pipes. When bi-metallic cells exist between connected pipes, the electrochemical current will attenuate (become smaller) away from the point where the two metals are connected. The attenuation is due to the increased electrolytic resistance to the farther ends of the pipe. The attenuation is a function of the cross-sectional area of the water in the pipe; however, it is expected that the current will fully attenuate within meters, if not centimeters or millimeters, of the union between the metals. Since the effects of current attenuation were not observed, it is suspected that polarization resistance associated with oxide film buildup minimized any galvanic interaction that may have occurred.

Major organic coating performance parameters central to the issue of corrugated steel pipe include resistance to:

- Water absorption.
- Water vapor transmission.
- Ultraviolet radiation.
- Impact-related disbondment and shattering.
- Abrasion.
- Underfilm water penetration/corrosion.

Assuming an organic coating is properly applied, four mechanisms limit its effective lifetime on CMP: mechanical damage, water absorption, water vapor transmission, and ultraviolet radiation. In areas where mechanical damage has not had an influence, water absorption and water vapor transmission may be the most likely mechanisms for coating deterioration. These mechanisms allow water to penetrate the coating and collect at the interface of the coating and substrate, negating the barrier protection of the coating. In most cases in the field, mechanical deterioration will be the major factor contributing to the deterioration of CMP coatings.

Mechanical damage includes impact-related disbondment, shattering, and abrasion of the coating. Impact-related damage may occur in handling, installation, or in service. Coating abrasion results from the bed load and water seen by the CMP. Because bed-load characteristics are transient factors that cannot be defined deterministically, mechanical deterioration will be a probabilistic factor.

Once mechanical influences have created coating defects exposing the substrate metal, corrosion and disbondment from those coating defects will control coating life. The ability of a coating to resist underfilm water penetration and subsequent corrosion is, therefore, a desirable property in CMP coatings.

Culvert Materials and Protective Coatings

This section discusses some alternative pipe materials and coatings that are available. This discussion does not purport to address all of the issues involved with each coating but is provided to help the highway engineer make informed decisions regarding relative durability when choosing a culvert material or coating. Most of the following discussions contain comments and conclusions made by other researchers from the literature review. As with all engineering materials, there are compromises involved in selecting culvert materials. Each material has its own strengths and weaknesses. It is not the intention of this section to make a case for or against any particular material or coating, but rather to make the highway engineer aware of the issues associated with various culvert materials and coatings.

Galvanized

Plain galvanized coating has been in service on corrugated steel pipe culverts since the early 1900's. Much research has focused on the performance of galvanized coating. It is a commonly used metallic coating and serves as the baseline for comparison for many studies.

In the manufacture of plain galvanized coated pipe, the steel sheet is zinc-coated through a hot-dip process in which the steel sheet is drawn through a molten bath of zinc. The zinc used for galvanizing is to be a minimum of 98 percent zinc. The minimum thickness of the coating is 2 oz/ft^2, or approximately 0.0043 cm (0.0017 in) of zinc per side of the steel sheet. The sheet is produced in one coating weight (thickness). Different thicknesses of corrugated steel pipe are obtained by use of thicker steel base materials. The minimum tensile strength of the galvanized sheet is 310 MPa (45 ksi), with a minimum yield strength of 228 MPa (33 ksi).

A study conducted by Caltrans (California DOT) subjected various coating specimens to abrasive bed loads using a rotating drum test.[4] The conclusion of this test was that galvanized performed better than Aluminized Type 2 or aluminum alloy, with Aluminized Type 2 a close second. Another study evaluated two Aluminized Type 2 field sites and seven galvanized field sites.[5] All of these sites are low-water stream crossings with moderate to severe corrosion and/or abrasion. The conclusion was that there may be a slight advantage with Aluminized Type 2.

Aluminized Type 2

The manufacture of Aluminized Type 2 involves the coating of steel sheet with commercially pure aluminum (referred to as Type 2) by a hot-dip process in which the steel sheet is drawn through a bath of molten aluminum. The minimum coating weight is 1 oz/ft^2, or approximately 48µm (1.9 mils) of aluminum per side of the steel sheet. The coating is only provided at one thickness designation. The minimum tensile and yield strength of the coated corrugated steel pipe material is 310 MPa (45 ksi) and 228 MPa (33 ksi), respectively.

A previous study conducted by the FHWA had a substantial portion of the analysis dedicated to the study of Aluminized Type 2 coating.[6] Using data from field inspections on 16 pipes coated with Aluminized Type 2, the study concluded that the coating was 6.2 times better than galvanized steel. This is based on comparison of actual Aluminized Type 2 field performance

44

with durability of galvanized as predicted by the California Method. A Florida DOT study concluded after a 5-year study by its corrosion laboratory that Aluminized Type 2 is 2.9 times better than galvanized in the same environment.[7] The *California Highway Design Manual* also claims that Aluminized Type 2 has durability advantages over galvanized.[8] It claims that 18-gauge Aluminized Type 2 will have a life equal to that of 16-gauge galvanized steel, but only between a pH of 5.5 and 8.5 and minimum resistivity of 3000 ΩZcm. Outside of these ranges, the design manual claims that galvanized and aluminized will show equal performance. Pyskadlo and Ewing state that Aluminized Type 2 test plate specimens performed better in field testing than did similar samples of Galvalume and galvanized.[9] In our own field investigations on the Santiam Highway, Aluminized Type 2 is outperforming the bituminous-coated and galvanized end sections of some culverts. But, on the Natchez Trace Parkway, two of the Aluminized Type 2 culverts showed accelerated corrosion. Aluminized Type 2 also displayed accelerated corrosion at the Dexter site in Maine.

From the literature review, it seemed that most researchers agree that in abrasive installations with considerable corrosion and abrasion, plain metallic coatings of any material are inadequate and additional coatings must be considered. Use of additional coatings may be a potential problem with aluminum alloy and aluminized culverts. In a study conducted by the New York DOT, it was found that polymer precoating showed less adhesion to Aluminized Type 2 coil steel than to galvanized coil steel.[10] In this study, test specimens with polymer-aluminum and polymer-galvanized coatings were subjected to weatherometer (ASTM G26), freeze thaw (ASTM A742), salt spray (ASTM B117), and immersion tests. The conclusion was that a polymer coating over aluminum is deficient and should not be used for New York DOT projects. Manufacturers of Aluminized Type 2 have contended that its performance in adverse environmental conditions (abrasion, poor pH, low resistivity) is so good that there is no need for additional coatings. The New York DOT investigation found 3 out of 23 culverts were experiencing "accelerated" corrosion (performing poorly). An additional barrier-type coating, if it could be applied, may improve durability at these locations.

Jacobs states that aluminum acts only as a physical barrier and provides inferior cathodic protection compared with zinc.[11] It is generally true that pure aluminum is a less efficient anode than zinc. This is why aluminum anodes have alloying constituents called "activators." However, the cathodic protection mechanism only works on exposed steel. One must consider that initially, the steel is mostly covered with the coating (zinc or aluminum). Until such time as a significant amount of steel is exposed, the coating is acting primarily as a physical barrier between the steel and the environment. Thus, the coating provides protection to the steel both as a barrier and through cathodic protection.

In the field studies of this investigation, a light coating of rust completely around the circumference of some Aluminized Type 2 pipes has been observed. The occurrence of this minor corrosion is referred to as "blushing" by the manufacturer and also referred to as "staining" by others in the industry. The resulting rust may be due to microfissures through the aluminum coating. These microfissures could be caused during the rolling process that forms the corrugations in the CSP. Obviously, the existence of rust in these fissures is an indication that the steel substrate is experiencing relatively insignificant corrosion. However, if no measurable thickness loss is observed any corrosion occurring would be insignificant. This phenomenon was observed in the field studies.

The damage to the aluminum coating during the rolling process is compounded at the ends of each section of pipe during the rerolling process. The pipe ends are re-rolled to provide positive locking with the connecting bands that hold the pipe sections together. This re-rolling overstresses the coating.[11] It has been observed in this field study and others that corrosion is more prevalent on these re-rolled sections. Applying a touch-up coating after re-rolling helps alleviate this problem. Coating damage is also a concern with the extreme bends performed during the lock-seam operation.

Aluminized Type 1

Aluminized Type 1 is a sheet metal coating that has recently been introduced for use on culverts. It is an aluminum coating with 5 to 11 percent silicon meeting ASTM A929. The coating is typically produced with a coating weight of 1 ounce per square foot of surface (total both sides) to result in a coating thickness of 48μm (1.9 mils) on each surface. The NCSPA *CSP Durability Guide* states that there is not enough field service history to provide add-on service life values.

Aluminum Alloy (Alclad)

Aluminum alloy pipe is fabricated from aluminum alloy sheet composed of a 3004 alloy aluminum core and a 7072 alloy aluminum cladding layer. The nominal cladding thickness is 5 percent of the total composite thickness, on each side of the sheet. Sheets are available at different thicknesses between 0.09 and 0.42 cm (0.036 and 0.164 in). The clad sheets are fabricated through cold-metal welding of sheets of 7072 alloy aluminum on opposite sides of a core sheet of 3004 aluminum during cold rolling. The sheets have a minimum ultimate tensile strength and yield strength of 214 and 165 MPa (31 and 24 ksi), respectively.

A study published by the Maine DOT in 1982 observed "exceptionally good durability for both fresh and salt water installations."[11] Aluminum alloy also proved itself in a New York DOT study conducted in 1987 where several 0.6-m by 1.2-m (2-ft by 4-ft) test plates with various coatings were installed in culverts. In that study, aluminum plates performed better than polymers, epoxy, Galvalume, and galvanized.[9]

There have been shipping, handling, and installation differences noted with the aluminum alloy. The relatively light weight of aluminum is an obvious shipping and handling advantage over steel. Although easier to handle, it has been reported that the light weight of the pipe gives it a tendency to float during installation, making installation more difficult.[12] Also, installation problems with aluminum alloy result because its strength is not comparable to steel. Having less strength, the pipes are more prone to damage during backfill operations. Stones against the culvert surface tend to puncture the surface, and improper compaction of the backfill material can cause defection in the pipe wall.

It was previously noted that when corrugating Aluminized Type 2 pipe, microfissures occur in the aluminum coating. This phenomenon has also been observed in aluminum alloy culverts. This was documented in a study published by the Indiana Department of Highways.[13] In an inspection of an aluminum culvert exposed to a very low pH environment, it was noted that "most of these fissures were orientated parallel to the corrugations and may be the result of stress induced corrosion or corrosion at the minute metal fractures induced at the time of fabrication of the pipe."

Stratfull is quoted by Malcom as stating, "At pH ranges 5-6 and 8-9 chemical stability of aluminum does not appear to be clearly defined as pH 6-8."[14] The report recommends that aluminum should always be coated when it will be exposed to pH in the ranges of 5 to 6 and 8 to 9. Young states that asphalt has poor adhesion to aluminum in comparison to galvanized and should not be used in corrosive or abrasive environments.[10]

Galvalume

In the manufacture of Galvalume corrugated metal pipe, the steel sheet is coated with an aluminum-zinc alloy by a continuous hot-dip process. The coating has a composition of 55 percent aluminum, 1.6 percent silicon, and 43.4 percent zinc. The minimum coating weight is 0.7 oz/ft^2 or 0.0028 cm (0.0011 in) per side of the steel sheet. The minimum tensile and yield strengths of the coated, corrugated steel material are 310 and 228 MPa (45 and 33 ksi), respectively.

In a field study conducted by the Maine DOT, an advantage for Galvalume over galvanized was observed in locations where both were installed.[11] This site was evaluated in the field studies conducted during this investigation.

One reference found that Galvalume outperformed both galvanized and Aluminized Type 2 in salt spray, cyclic standing, and full immersion tests.[15] In the same study, it was also noted that in field tests, Galvalume was doing better than galvanized at 16 sites of 9- to 9.5-year-old culverts.

Bituminous-Coated

Bituminous (asphalt) coating can be applied to either metallic-coated steel or aluminum pipe materials. The coating may cover different areas of the pipe for use in different environments. There are four methods of applying bituminous coating to satisfy various installation environments. The first is designated as Type A coated pipe and has a uniform coating of asphalt on the interior and exterior of the pipe, with a minimum thickness of 0.13 cm (0.05 in). Type B has a uniform coating on the inside and outside bottom 180 degrees of the pipe to a minimum thickness of 0.13 cm (0.05 in). The bituminous material fills the corrugations of at least the bottom 90 degrees of the interior of the pipe, providing a smooth invert. The pavement has a minimum thickness of 0.32 cm (0.125 in) above the crest of the corrugation. Type C is fully coated as with Type A, and has a smooth invert, as in Type B. Type D is fully coated as is required with Type A, and has a thick coating centrifugally spun inside the entire surface of the interior of the pipe, filling the corrugations completely around the full inside circumference of the pipe. The lining has a minimum thickness of 0.32 cm (0.125 in) above the crest of the corrugations.

The bituminous material is made of asphalt. It is soluble in trichloroethylene, and must pass shock, flow, and imperviousness tests. There are difficulties in handling and installing bituminous-coated CMP because of its tacky coating and increased weight.

Bituminous coating is applied by a hot-dipping process. The process itself has also been scrutinized for leaving imperfections in the coating. It has been observed that these imperfections allow interaction between the CMP and whatever effluent the pipe is carrying.[13] Storage, often necessary after manufacture, has been found to cause damage. Bitumen is susceptible to ultraviolet degradation.[13,16]

A 1959 California study concluded that bituminous coating is only useful as a moisture barrier and suffers from accelerated deterioration under abrasive flow.[17] This study went on to conclude from field data that bituminous coating does increase service life by an average of 6 years in a corrosive flow. Georgia and Florida studies have data indicating additional service life from bituminous coating as 10 years for low-velocity, non-abrasive flows.[1, 3] In the field studies presented in this report, bituminous-coated galvanized pipe was experiencing accelerated deterioration on installations with abrasion. The *California Highway Design Manual* provides criteria for bituminous-coated pipe.[8] Table 11, copied from the manual, is provided as a guide.

Table 11. Anticipated service life added by protective coatings.[8]

Flow Velocity	Channel Materials	Bituminous Coating (years)	Bituminous Coating and Paved Invert (years)
<1.5 m/s [<5 fps]	Abrasive	6	15
	Non-Abrasive	8	15
1.5-2.1 m/s [5-7 fps]	Abrasive	6	12
	Non-Abrasive	8	15
>2.1 m/s [>7 fps]	Abrasive	0	5
	Non-Abrasive	2	10

The *California Highway Design Manual* goes on to state that bituminous coating acts as a good inhibitor for soil-side corrosion. This is of particular interest for arid regions where soil-side corrosion may be dominant. AISI says that under average conditions, bituminous coating will add 25 years to service life when applied to the soil side.[18]

Polymer-Precoated

Galvanized and aluminum-zinc alloy-coated steel pipe are available with a polymer coating on one or both sides (Aluminized Type 2 with polymer coating was not commercially produced at the time of this report). The polymer precoat material was one of the promising materials for metal CSP protection evaluated in the NCSPA study on durability of culvert coatings.[23] Different coating grades are available and are based on the thicknesses of the coating on the two sides of the pipe. Grades have included 10/0, 10/3, and 10/10, which correspond to 0.010 in/0 in (0.025 cm/0 cm), 0.010 in/0.003 in (0.025 cm/0.0076 cm), and 0.010 in/0.010 in (0.025 cm/0.025 cm) thicknesses of coating on the inside/outside of the pipe. Recently, 10/10 has been the only polymer coating commonly available. The polymer coating must pass adhesion, impact, thickness, and holiday tests. Abrasion resistance, imperviousness, freeze-thaw resistance, weatherability, and resistance to microbial attack tests may also be included for specific applications. The type of polymer is not specified.

Polymer precoat is a mill-applied coating. This means that the coating is applied to the sheet metal before the corrugation is performed. One problem that has been found with this is that stress involved with making the extreme bends of the lockseams weakens the bond between the coating and the steel.[19] It has been observed that this damage can cause the coating to pull away under extreme conditions (i.e., heavy abrasive, low pH).

Polymer precoat shows better durability than bituminous coating. One study noted that the as-received samples of polymer-precoated pipe showed considerably less damage, were easier to handle, and were 1/3 the weight of bituminous-coated pipe.[16] It was also observed in the study that after a year of exposure outdoors, there was no evidence of chalking, hardening, or other damage due to aging on the polymer precoat specimens. After evaluating samples with bituminous and polymer precoat under different abrasive bed loads at varying temperatures, the same study stated that the polymer precoat samples evaluated were superior to the bituminous-coated samples.

New Technology Coatings

In one study, laboratory evaluations of epoxy-coated, polymer precoated, bituminous-coated, Galvalume, aluminized, and other CMP exposed to extremely low pH discharges showed epoxy coating to be the best.[13] Epoxy coating also proved its value in a 1987 study conducted by the New York DOT. In this study, several 0.6-m by 1.2-m (2-ft by 4-ft) test plates with various coatings were installed in culverts. "Epoxy plates performed better than all the polymer plates after 2 years exposure in an aggressive site," except where boulders were in the bed load, then polymer plates were better.[9] A possible explanation of this could be that the epoxy coating is showing better abrasion resistance than the polymer coatings, but since epoxy is a brittle coating it is more susceptible to impact damage.

The NCSPA is currently conducting a multi-phase program with the objective of developing test procedures that will be used to develop improved coatings for protection of the invert of corrugated steel pipe. The program has resulted in a four-tier test protocol for evaluation of new CSP coating materials. Central to this protocol are tests that focus on the abrasion resistance of the novel materials as applied on actual pipes. This test protocol is available to manufacturers interested in specifying their product for use on corrugated steel pipe.[23]

One of the newest technologies to be evaluated includes polymer-modified asphalt. These materials are similar to traditional asphalt coatings in their application, but are modified with polymer additives to provide improved durability.

Concrete

Concrete and reinforced concrete pipes are composed of cement, aggregates, and possibly reinforcing material. Concrete and reinforced concrete pipes are available in circular, arch, and elliptical shapes. They can also be provided to specific strength levels, as measured by a three-edge bearing test.

Portland cement itself is made of essentially hydraulic calcium silicates, which may contain a form of calcium sulfate. Portland cement concrete typically has an air-entraining addition mixed in the material, but not in reinforced concrete pipes. Types of cement with moderate and high sulfate resistance and low heat of hydration are also available. Fly ash and raw or calcined natural Pozzolan (siliceous or siliceous and aluminous materials) may be used as mineral admixtures in the portland cement concrete.

Fine and/or coarse aggregates may be added. Fine aggregate consists of natural or manufactured sand. The aggregates are to conform to grading criteria. A small amount of deleterious substances are allowed to various small mass percentages (clay, coal, extremely fine material, other concrete, shale, etc.). Coarse aggregate consists of gravel, crushed gravel, crushed stone, crushed air-cooled blast furnace slag, crushed concrete, or a combination of any.

Jacobs predicts service life between 65 and 70 years from statistical evaluation of concrete culverts.[11] In an Indiana study, evaluations of pipe under a severely acidic environment (pH 2 to 4) showed that concrete pipe was not suitable in this environment because of accelerated softening of the mortar.[13] It was also observed that when the softened mortar was not removed from the invert, the softening would limit itself.

Plastic

Two types of commonly used plastic pipe are polyethylene (PE) and polyvinyl chloride (PVC). Corrugated polyethylene pipe is made of virgin or reworked PE compounds through pipe extrusion or rotational molding. Pipes have an inside diameter between 30 and 91 cm (12 and 36 in) [7.6 to 25 cm (3 to 10 in) is considered tubing]. In contrast, metal pipes have inner diameters between 10 and 305 cm (4 and 120 in). Plastic pipe must pass several tests, including flattening, elongation, stiffness, environmental stress cracking, high-temperature strength, and low-temperature flexibility. A pipe made of polyethylene has a minimum stiffness between 152 and 310 kPa (22 and 45 psi) at 5 percent deflection for various diameters.

PVC pipes come in diameters from 46 to 122 cm (18 to 48 in), and can be used in unpressurized storm drains, culverts, and other sub-surface drainage systems. The pipe wall thickness shall be a minimum of 0.241 to 0.483 cm (0.095 to 0.190 in) with a minimum pipe stiffness of 83 to 220 kPa (12 to 32 psi), depending on the pipe diameter. The pipe shall undergo flattening, impact resistance, stiffness, and acetone immersion testing.

Weight is an advantage for plastic pipe. Because of its light weight, plastic pipe is cheaper and easier to ship, handle, and install. Another advantage of plastic pipe during backfill is that it can be cut with a conventional tool like a handsaw. But, like aluminum alloy pipes, problems arise with backfill operations. Extreme care must be taken not to damage the pipe during backfill. Rocks placed too close to the surface can puncture the pipe, and backfill compacting can cause deflections to occur in the pipe.

Vitrified Clay

Extra strength, vitrified clay pipe is manufactured from fire clay, shale, surface clay, or a combination of these materials that are combined and fired. Pipes can be between 7.6 and 107 cm (3 and 42 in), with a three-edge bearing strength of (2,000 to 7,000 lbf/linear ft), depending on the pipe diameter. Pipes may or may not be supplied with a glaze coating. Pipes must pass bearing strength, absorption, and acid resistance tests.

In a study conducted by the Utah State Department of Highways, all metal sections in a test were affected while vitreous clay was not affected.[20] Vitrified clay is reported to be durable even in highly acidic conditions with a pH between 2 and 4.[13] In one field study, 14 vitrified clay-lined culverts were in excellent condition after 3 to 12 years of exposure in extremely acidic conditions in Ohio.[19]

Durability Estimation Procedures and Examples

This section discusses guidelines for selecting culvert materials. Where possible, examples of the methods are given on the basis of field testing locations in this study. The first three methods discussed (California, AISI, and Florida) provide life predictions for the culvert materials. The remaining three methods (NCSPA, New York, and Colorado) are more focused on material selection than life prediction.

California

The California Method is based on the field studies of 7,000 culverts in the 1950's. Beaton and Stratfull discuss the development of the original procedure.[17] The researchers collected the data and performed a graphical analysis of environmental factors. By trial and error, the best parameters were selected for correlation with service life. The final product was a graphical portrayal of the relationship between pH and resistivity to years to perforation of galvanized steel pipe. This test method has been further refined over the years as a California DOT standardized test procedure, California
Test 643. The most recent updates to this method were made in 1993. Figure 18 shows the graph, taken from California Test 643, predicting years to first perforation for a 16-gauge CSP with a 56.7-gm (2-oz) zinc coating. Figure 18 was taken from the 1993 revised test method.

The chart in figure 18 is a relatively simple tool to use. To get input for the chart, the engineer must have pH and resistivity data for soil and/or water. The worst pH and minimum resistivity are used as inputs to calculate the life of a 16-gauge galvanized culvert. Equation factors are provided as an alternative to the graphical solution. Multipliers are given to adjust the life prediction for gauges other than 16.

The California Method is now the most widely accepted method to estimate culvert durability. The researchers of the original test method recognized the possible shortcomings of the test method in the 1959 study. They go on to say that accuracy of the test method will increase when more corrosion variables are considered. Other factors (e.g., frequency of flow, abrasion, bacterial action) are inherently in the test method as statistical averages of the randomly sampled culverts that form the basis of the study.[17] The California Method will not be suitable for predictions of durability of culverts in extreme conditions since it is an average by definition.

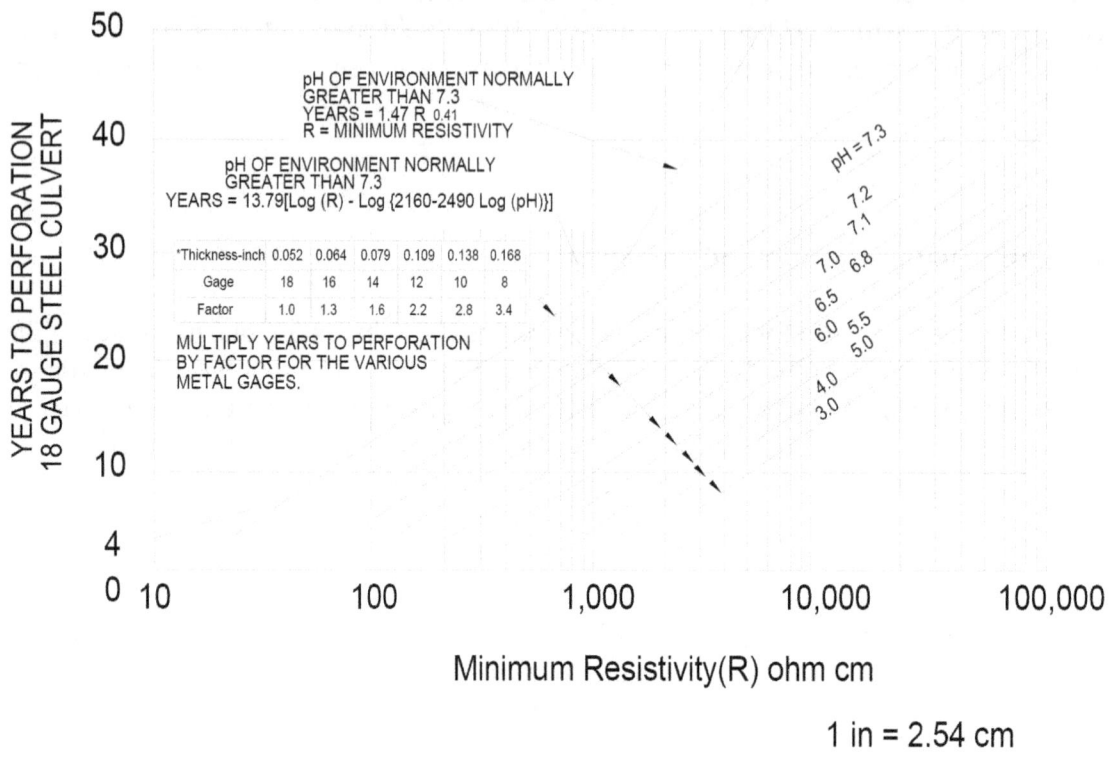

Figure 18. California Method chart for estimating years to perforation.

Table 12 contains a summary of evaluations of the California Method. These evaluations were taken from studies published by various researchers that were uncovered in the literature review. The table suggests that there is some degree of variance between the California Predictions and the observed field performance of galvanized steel culverts. This method serves more as a guide to durability than as a measure of exactly how long the culvert will last. A standard error of ±12 years is noted in the original study. One contribution to the scatter in the data may be that California is made up of many climatic regions, which may have had differing effects on the culverts that were part of the study.

Table 12. Selected research conclusions about the California Method.

Reference	Conclusions about the California Method on the basis of data and/or observations
Florida Reference 7	Accepts the California Method as suitable for the performance of galvanized in the Florida environment but develops new equations to predict durability for Aluminized Type 2, aluminum alloy, and concrete.
Oklahoma Reference 21, pp. 9	The California Method generally does not correlate with the observed culvert conditions in the State. The method predicts a shorter lifetime than observed in the western two-thirds of the State, with the exception of the high plains area of the panhandle where it was quite accurate.
Georgia Reference 3, p. 6.1	On the basis of a survey of 251 culverts (140 plain galvanized) in Georgia, it was concluded that expected service life was 50 percent greater than that predicted by the California Method. The AISI method is consistent to conservative in Georgia.
Louisiana Reference 12, p. 32	"Under the environmental conditions (moderately to very corrosive) encountered during this study, the California Chart overestimates predicted pipe life. The chart does, however, combine pH and resistivities to correctly predict life in a relative sense for the mildly, moderately, and very corrosive environments."
Idaho Reference 2, p. 3	"The test developed by the California Division of Highways and their service life chart appears to be satisfactory. It appears the test method estimates the service life conservatively in all but a few installations."

American Iron and Steel Institute (AISI)

The AISI method is very similar to the California Method.[18] Figure 19 shows the AISI chart. It is graphically similar to the California chart except for the service life values. The California Method assumes maintenance-free service life is the number of years of service until the culvert obtains its first perforation. AISI takes the position that 13 percent average metal loss occurs in the invert at first perforation and that 25 percent metal loss in the invert marks the limit of a culvert's service life. Therefore, AISI predictions are twice that of the California Method. In a field evaluation of 140 plain galvanized culverts in Georgia, it was found that the AISI predictions were accurate, but the author of that study recognized that the AISI predictions are generally conservative.[13] The AISI chart contains multipliers for different gauges as does the California chart.

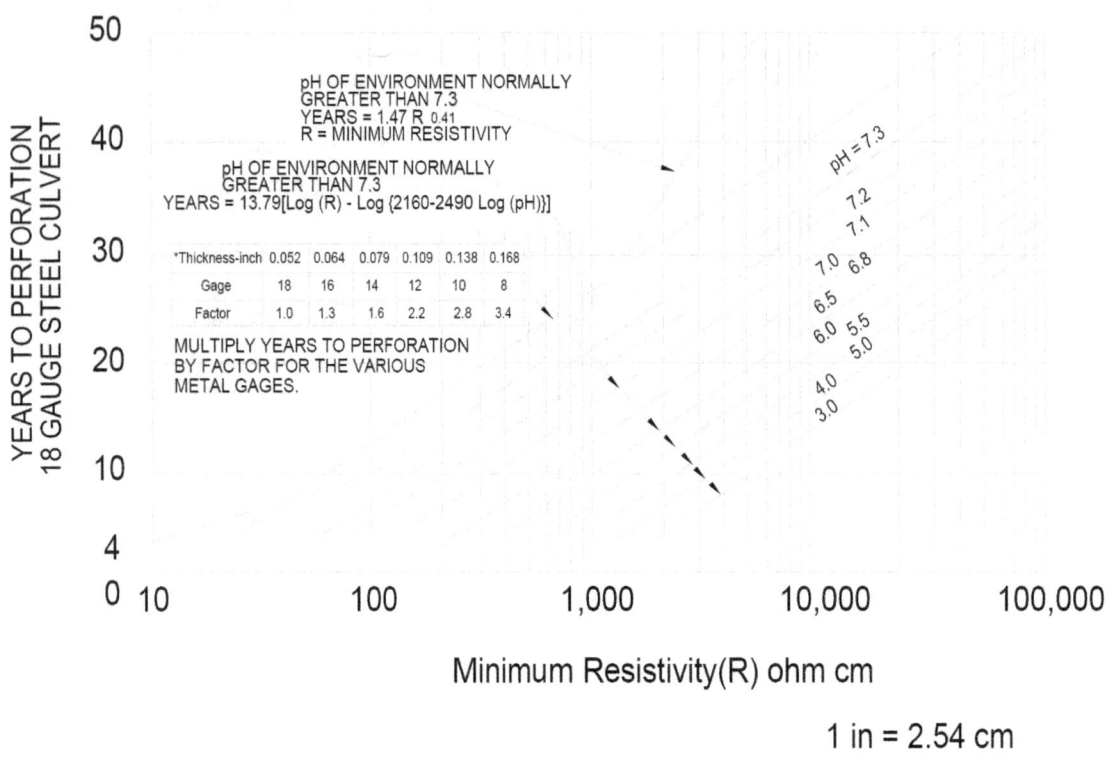

CHART FOR ESTIMATING YEARS TO
PERFORATION OF STEEL CULVERTS

Figure 19. AISI chart for estimating years to perforation.[18]

Florida

A study published in 1993 by the Florida DOT presents a durability prediction method for Aluminized Type 2-coated corrugated steel, concrete, and aluminum alloy culverts.[7] The method developed is derived from the California Method. A study conducted by the Florida DOT Corrosion Laboratory concluded that the life of Aluminized Type 2 coating is 2.9 times that of galvanized. Using their performance data, the findings of FHWA-FLP-91-006, and the California Method, Florida DOT came up with the service life estimation graph shown in figure 20.

The environmental data forming the basis of the predictions were from field studies conducted by a Florida culvert company. Detailed site descriptions were not provided in the study.

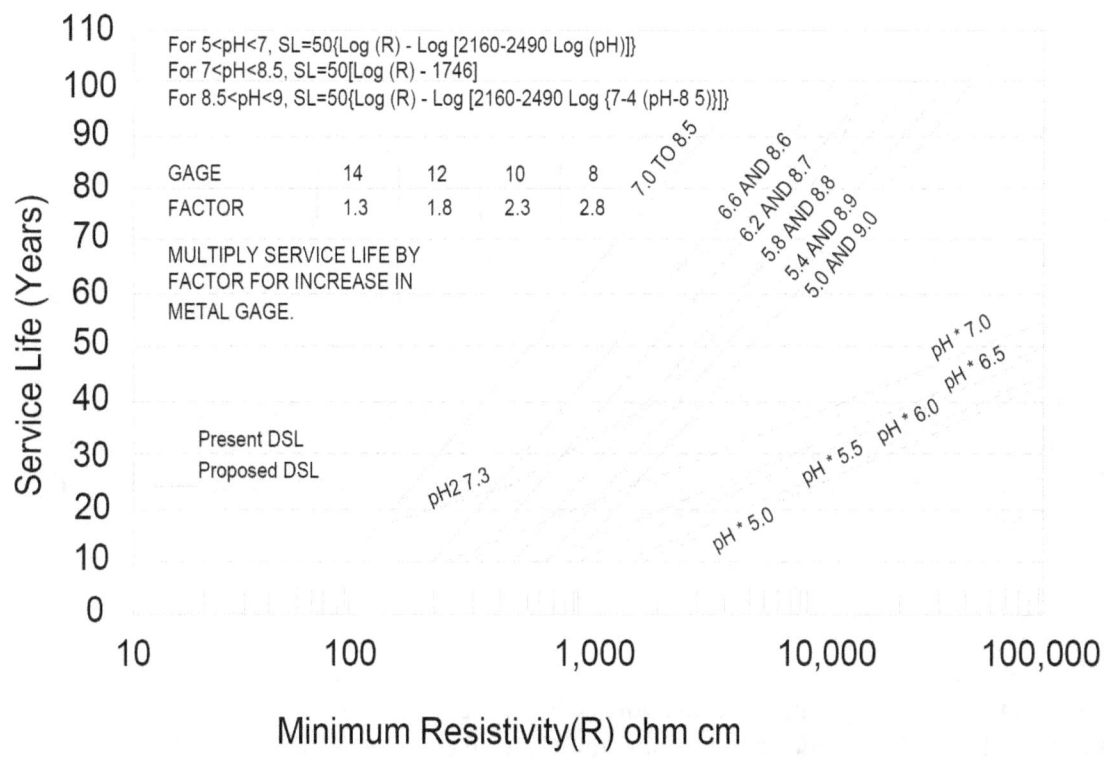

Figure 20. Florida DOT chart for estimating
years to perforation of Aluminized Type 2.

National Corrugated Steel Pipe Association (NCSPA)

The NCSPA *CSP Durability Guide* is a product usage guide for corrugate steel pipe. The four-page document provides usage guidelines for a variety of corrugated steel pipe products. Fourteen coatings are represented in the guide. Products are positioned relative to corrosion condition, abrasion level, and environmental characteristics. The guide includes the AISI chart for predicting the service life of CSP and provides a table of add-on service lives for various coatings. The method is intended to be a guide to be used by the highway engineer to help in the selection of culvert materials and coatings. It is not a substitute for professional engineering advice for specific applications. The NCSPA *CSP Durability Guide* is reproduced in Appendix C.

New York

This method describes the galvanized steel pipe criteria used by the New York State DOT. The method includes five criteria to determine the "durability index" of the culvert location. The criteria include:

- Geographical area (rating of 1, 3, 5, 7, or 9) - generated by dividing the state of New York into geographical areas, which are rated in five categories depending on relative soil corrosiveness.

- Abrasion (rating of 1, 2, or 5) - depending on bed load, gradient, and "relative abrasiveness."

55

- Flow condition (rating of 1, 2, or 3) - highly intermittent, intermittent, or continuous.

- Service rating (1 or 2) - side drains or driveway pipes versus cross-drains.

The total of the ratings for a given location is the "durability index." A recommendation on whether to use a paved (bituminous) or unpaved culvert is based on this index value. Index values above 13 recommend paved inverts; values 13 and below recommend plain galvanized product.

The method does not have a graphical figure for predicting service life; rather, it uses a numerical system based largely on experience to recommend appropriate materials. This rather simple approach to determining the appropriate material for a culvert is perhaps the most typical of those used by State agencies.

Colorado

This method was distributed by the Department of Highways in Colorado.[22] As with the New York method, this method was included to address the culvert material selection concerns that exist at the State level. It is intended to provide guidelines for culvert installations, but Colorado highway engineers are not required to follow these guidelines.

This guideline provides an index that uses inputs of sulfate, chloride, and pH measurements of both soil and water to calculate a corrosion rating (CR) of 1 through 6. Table 13 is used to establish a CR for the installation site. Once a CR has been established from table 13, table 14 is used to select the pipe material.

Table 13. Corrosion rating (CR) table used by Colorado DOT.

CR Level	Soil Characteristics			Water Characteristics		
	Sulfate (SO_4) % max	Chloride (Cl) % max	PH	Sulfate (SO_4) Ppm max	Chloride (Cl) ppm max	PH
CR 0*	0.05	0.05	6.0 – 8.5	250	250	6.0 - 8.5
CR 1	0.15	0.15	6.0 – 8.5	250	250	6.0 - 8.5
CR 2	0.05	0.05	6.0 – 8.5	500	500	6.0 - 8.5
CR 3	0.15	0.15	6.0 – 8.5	500	500	6.0 - 8.5
CR 4	0.50	1.00	5.0 - 9.0	1000	1000	5.0 - 9.0
CR 5	1.00	1.50	5.0 - 9.0	2000	2000	5.0 - 9.0
CR 6	>1.00	>1.50	<5.0 or >9.0	>2000	>2000	<5.0 or >9.0

* No special corrosion protection recommended when values are within these limits.

Table 14. Pipe material selection table used by Colorado DOT.

Corrosion Resistance Number	CR 1	CR 2	CR 3	CR 4	CR 5	CR 6
Corrosion Condition Description	Mild	Mild	Mild	Moderate	Severe	Extreme
Corrosion Condition Inside or Outside Pipe	Outside Only	Inside Only	Both	Both	Both	Both
Type of Pipe						
CSP	No	No	No	No	No	No
Bit. Co. CSP	Yes	No	No	No	No	No
ASB Bo. CSP	Yes	Yes	Yes	Yes	Yes	Yes
CAP	Yes	Yes	Yes	Yes	Yes	No
PCSP Both Sides	Yes	Yes	Yes	No	No	No
PCSP Outside	Yes	No	No	No	No	No
PCSP Inside	No	Yes	No	No	No	No
RCP or NRCP, Type I Cem.	Yes	Yes	Yes	No	No	No
RCP or NRCP, Type II Cem.	Yes	Yes	Yes	Yes	No	No
RCP or NRCP, Type IV Cem.	Yes	Yes	Yes	Yes	Yes	Yes

PCSP - Precoated Corrugated Steel Pipe
NRCP - Unreinforced Concrete Pipe

Discussion

It is clear from the review of literature and the research conducted in this study that no one method considers all of the factors affecting culvert durability or precisely predicts how long a culvert will last. The California Method appears to be a pragmatic tool for estimating culvert durability in the design stage. The method does seem to be limited in severe environments where pH, resistivity, or abrasion become a serious factor. It is necessary to quantify the environment in which a culvert is to be installed. Once quantified, the predictions from the California Method can be used with caution. Comparison against culverts used in the same or similar environments should be made to verify the predictions.

A more precise method of durability prediction needs to be developed. Because of the limited number of samples in this study, it was not possible to develop a durability method within the scope of the contract. The closeness in age and environmental conditions for the culverts in this study would have made any such method of limited value. But, a database of galvanized pipe condition was developed from the literature review. This included data on 240 galvanized culverts in three States. It would be helpful if future research could provide additional data for this database; however, the lack of consistency in data acquisition and evaluation makes comparison difficult.

Once enough data have been collected, statistical methods can be employed to develop predictive methods based on the most significant criteria. In the past, resistivity, pH, and percent perforation have been considered key factors. In addition to these factors, data are required on the abrasive potential of the site. During this investigation, a relationship was observed between abrasion potential and pipe deterioration. Quantifying abrasion potential is difficult, thus it is not uniformly evaluated in various studies. Work is needed to understand and quantify abrasion potential.

Software for Estimation of Durability

CULVERT 3 - Caltrans

CULVERT 3 software calculates the service life of a culvert on the basis of the following California Standard Methods: 417, 422, 532, 643 and the *California Highway Design Manual*, Section 850. Methods 417 and 422 are standard methods for testing soils and waters for sulfate and chloride content. The *California Highway Design Manual's* Section 850 deals with the selection of drainage material size and type. Standard test method 643 is the well-known and widely used California Method. The software combines these sampling and predicting techniques in an attempt to improve upon the original method. One major improvement on the original method is the consideration of abrasion and flow velocity. The program inputs values for the flow velocity and the abrasive character of the existing site. It is uncertain exactly how these variables are considered; abrasion and flow are not considered in Test Method 643. The software included with this report is based on Method 643, not the method of CULVERT 3.

This program begins by prompting the user for various descriptive information about the culvert in question, including location, date, test number, etc. Next the program prompts for the pH and resistivity of either the water or soil or both. Similar to the original method (Standard Method 643) it is only necessary to provide data on the soil or water. If both soil and water parameters are given, the program will automatically select the worst case. The next two questions allow the user to correct for stream flow and abrasiveness. One of the prompts asks if you would simply like to cut the service life estimate in half on the basis of the severity of the flow. If you choose to correct for the stream flow and abrasiveness, the next screen provides a chart. The chart allows the user to choose one of eight flow situations. The choices are such that abrasion is a function of the flow velocity. After the flow correction is selected, the user is given the life prediction results.

During this investigation, it was found that abrasive potential of the site was more important than the flow. An example of this is the difference between the Santiam sites and the Natchez Trace Parkway sites. At the Santiam sites, the flow velocity was considerably higher on average than the Natchez Trace Parkway sites, but the Natchez Trace Parkway sites showed more evidence of corrosion than did the Santiam sites.

If the environment is within the limits of the California Method, the results of the program are in the form of a table of life predictions that displays varying alternatives of galvanized, bituminous, and bituminous and paved pipe along with varying gauges. Along with the table, recommendations for alternative materials and coatings (reinforced concrete pipe [RCP], aluminum, etc.) are provided. For certain extreme environments (e.g., when the resistivity is less than 1000 ohm-cm), the California standards do not recommend the use of metal pipe and instead recommend the use of RCP. If the environment is not favorable for metal pipe, the message "Due to aggressive environment the use of corrugated metal pipe is not recommended," is displayed followed by a prompt for chloride and sulphate data. California Standard 532, "Method for Estimating the Time to Corrosion of Reinforced Concrete Substructures" requires data for chlorides and sulphate in order to make a service prediction for RCP.

Culvert Service Life Estimator (CSLE) - Florida DOT

The Culvert Service Life Estimator (CSLE) was developed by the Florida DOT Corrosion Research Laboratory. CSLE bases the logic of its culvert service life predictions on the California Method. The Florida DOT Corrosion Laboratory used its own field data compiled over 3 years to adjust or "custom-fit" the California equations to estimate the service life of galvanized, Aluminized Type 2, Alclad, and concrete pipes. The predictions of service life for galvanized are the same as those given by the California Method. For the alternative materials of Aluminized Type 2, Alclad, and concrete, the program calculates a factor that was derived by the Florida DOT Corrosion Laboratory from field investigations of those materials.

The program provides similar results to the California Method and provides correction factors for the alternative materials. The program does not consider abrasive flow.

Plain Galvanized Exterior Corrugated Pipe Life Prediction - NCSPA

This software is limited to the consideration of soil-side corrosion. This computer program was developed to predict the average service life for sites where durability is limited by soil-side corrosion. The program requires measurement of soil moisture content, soil resistivity, pH, and chloride ion concentration. Data used to develop this relationship came from field observation of 162 culverts.

This program is based on the percentage of zinc remaining on the culvert. On the basis of the input variables, the program predicts the condition of the protection over time. When the galvanized coating reaches the point where pitting of the steel substrate could theoretically begin, the program models further corrosion as if the culvert were black steel. The black steel model is taken from existing data on 23,000 black steel underground storage tanks.

The program compares the local soil corrosivity with the field data from 162 culverts. On the basis of its similarity, the probability of bare steel exposure is determined. Provided with this probability, the model assumes a steel corrosion rate and determines a probable time to failure.

APPENDIX A. REVIEW OF PERTINENT REFERENCES

The body of this report discusses a number of critical issues relating to culvert durability. The references cited form the basis of the majority of this discussion. This appendix provides, in tabular form, a listing of selected references along with brief comments on the contents of the reference. This is not intended to be comprehensive, but rather to assist the reader who wishes to pursue additional references. The references below are listed chronologically and are cross-referenced to the cited references.

Reference	Comments
Highway Design Manual, Section 850-13, California DOT. ref. 8	This highway design manual suggests a specific pH and resistivity range where Aluminized Type 2 CSP will outperform galvanized. Beyond this range, they will perform similarly.
"Method of Estimating Corrosion of Highway Culverts by Means of Polarization Cures," R.I. Lindberg, Research Scientist, Reynolds Metals Company, Not dated.	This paper covers development of a method for estimating corrosion by means of polarization curves. The investigator uses a method of the National Bureau of Standards and applies it to full-size metal culverts. The method involves obtaining curves by using four auxiliary anodes symmetrically placed near the ends of the culvert and two reference electrodes placed over the centerline of the culvert.
"Underground Corrosion," Melvin Romanoff, National Bureau of Standards, 1957.	This paper is the final report on studies of underground corrosion conducted between 1910 and 1955 by the National Bureau of Standards (NBS). Until 1922, corrosion research was mainly in the area of stray current electrolysis. These studies performed by NBS stem from the discovery that serious corrosion occurs in soils under conditions that preclude stray currents. The studies discussed in this paper are extensive investigations into the rate of corrosion in an underground environment of many alloys, both coated and uncoated.
"Field Test for Estimating the Service Life of Corrugated Metal Pipe Culverts," State of California Department of Public Works, J.L. Beaton, R.F. Stratfull, January 1962. ref. 17	This paper discusses the development of the "California Method." This method is the result of investigations performed over a 35-year period, during which 12,000 corrugated pipes have been evaluated in California. Procedures are presented on survey and sampling techniques.
1962 Reinspection of Coated and Paved Corrugated Steel Pipe in Flooded Control Levees, St. Louis District, Corps of Engineers, Armco, 1962.	This report includes a discussion of the qualitative performance of coated and paved corrugated steel pipe used in levees. The investigation concluded that all pipes were in such good condition after 29 years that it was not yet possible to determine the difference in service life between plain galvanized and paved and asbestos-bonded and paved.
Corrosion of Steel Pilings in Soils, U.S. Department of Commerce, 1962.	This report provides results from inspection of steel pilings. Excavations to depths of 4.6 m (15 ft) were made and steel pilings that have been in service for 7 to 25 years were inspected. A literature survey of pertinent topics was conducted. Various environmental conditions were recorded at excavation sites, including water chemistry, pH, and resistivity (some of these measurements were taken at different depths for the same site). This study is not directly related to culverts, but the data concerning depth in relation to different corrosion rates could be useful.

Report on Inspection of Bituminous-Coated and Uncoated Galvanized Metal Culvert Pipe, Department of Commerce, October 1964.	This report concerns an inspection of bituminous-coated paved culverts and plain galvanized culverts. Both coatings were approximately 30 years old at the time of this inspection. Various environmental factors and dimensions were recorded. Conditions and corrosion ratings are given for the current inspection and an inspection that was conducted 10 years prior.
"Mississippi Pipe Evaluation Study," Russell Brown, November 1964.	This study was a field inspection of 785 pipe structures, both metal and concrete. Mostly qualitative data were recorded at each site. Visual standards based on percentage of coating deterioration of each structure were used in structural and material ratings.
Soil Resistivity as Related to Underground Corrosion and Cathodic Protection, William J. Schwerdtfeger, March 1965.	This report is a discussion of data collected by the National Bureau of Standards on 4,500 specimens of commonly used ferrous materials. The specimens were buried in backfilled trenches for up to 17 years in soils that ranged in resistivity from 50 to 54,000 ohm-cm and in pH from 2.6 to 10.2. From the data, average maximum pit depth was extrapolated and weight loss was converted to corrosion current density.
"Durability of Metal Pipe Culverts," State of Idaho, Department of Highways, 1965. ref. 2	This study concerns a field and laboratory investigation into the performance of galvanized metal pipe. The California Method was used to predict the life of the culverts in this evaluation at the time of their installation. The author of this study reevaluates the culverts after 7 years and concludes that the California Method is accurate as a tool for durability predictions. Galvanized steel was the focus of this investigation.
"Durability of Corrugated Metal Culverts," State of New York, DOT, November 1967.	This paper presents results of two different studies. The first was a statewide survey of approximately 800 coated, uncoated, and coated/paved galvanized steel culverts ranging in age. The second is an interim evaluation of aluminum and galvanized steel culverts exposed to similar environments.
Durability Design Method for Galvanized Steel Pipe in Iowa, W.J. Malcom, Spring 1968. ref. 14	This report is a presentation of a durability design method. The design method is derived from field durability investigations that determined the actual service life of plain galvanized pipe in various parts of Iowa. Three soil groups in Iowa varying from acidic soil to alkaline soil were investigated.
"Abrasion Tests on Several Types of Metal Culvert Pipe," American Railroad Engineers Association, 1968.	This paper provides the results of a test following a procedure described in the American Railroad Engineers Association (AREA) Specifications for Bituminous-Coated Culverts. The pipe sample being tested is closed at the ends with wooden bulkhead and revolved end over end about a transverse axis at a speed of approximately 4 revolutions/min in such a manner that the erosive charge contained in the pipe rolls along the inner surface of opposite sides of the pipe. The author discusses the results in detail. Various coatings were tested and it was found that the asphalt coating stood up well and the epoxy coating failed quickly.
"Corrosion and Abrasion Design," Presented at AISI-NCSPA Storm Sewer Seminars, 1968-1972, William C. Clements.	This paper is a discussion of the factors affecting CMP corrosion and the methods of durability predictions used by various agencies. No data are presented.
Alabama – Detrimental Effects of Natural Soil and Water Elements on Drainage Pipe Structures, Luther W. Hyde, August 1969.	This report discusses field studies evaluating soil and water parameters and their relationship to culvert pipe corrosion. Various coatings and environmental conditions were considered. From results a graphical method based on pH, resistivity, and dissolved oxygen content was devised for estimating the life expectancy of steel pipe.

"Comparative Study of Coatings on Corrugated Metal Culvert Pipe," March 15, 1971. ref. 16	This paper is a comparative study of various CMP coatings under abrasive, corrosive, and other environments (i.e., freeze/thaw, high/low pH). This is a laboratory investigation only; there are no field data presented. Abrasion bed load is discussed in detail, including material, hardness, and sieve analysis.
"Durability of Corrugated Steel Culvert in Oklahoma," Oklahoma Department of Highways, C. J. Hayes, 1971. ref. 21	The purpose of this study was to determine the corrosiveness of Oklahoma's soil and water. A corrosion stress map of Oklahoma was created from the data. The evaluation included testing of 2000 soil samples and the inspection of 800 steel culverts. Included is a discussion of studies done by other investigators. Much field data on pH and resistivity are presented for various locations in Oklahoma. The author concluded on the basis of the data that the California Chart generally lacks correlation with observed performance.
"Steel Products for Culvert Applications," Presented at 1972 Highway Research Board Meeting, Washington, DC, J.D. Swan, U.S. Steel Corp., 1972.	This is a two-part field investigation into the performance of various galvanized steel culverts and an evaluation of culvert materials with a corrosion and abrasion resistance superior to that of galvanized steel. A considerable number of coatings and environmental conditions were considered. The results of this study state that pot-dip galvanized has the same corrosion resistance as continuous-galvanized.
"Evaluation of Aluminum Alloy Pipe for Use in Utah's Highways," Utah State Department of Highways, July 1973. ref. 20	This paper is an investigation into the feasibility of aluminum alloy pipe as an alternative to galvanized steel in Utah State Highway projects. Evaluation of eight pipe sections in high-alkalinity environments was performed. A survey of many State agency practices regarding aluminum alloy pipe was conducted. Study results stated that aluminum alloy pipe showed good performance after 7 years in alkaline soils.
"Pipe Corrosion and Protection Coatings," Utah State Department of Highways, November 1974.	This was a durability study of pipe culverts. Results drawn in this study are for the alkaline soils of Utah and do not propose to address all environmental conditions. This investigation included the inspection of 58 pipe culverts with graphical representation of data. The author presents a thorough analysis of many environmental factors and develops equations for durability predictions.
"Performance Evaluation of Corrugated Metal Culverts in Florida," R.P. Brown and R.J. Kessler, November 1975. ref. 1	The purpose of this study was to establish a correlation between culvert performance and environmental conditions as a function of time. A field survey of 153 culvert installations in Florida was conducted. Performance of culverts was evaluated against the California Method, AISI Method, and the New York DOT Index Method.
"Aluminum Culvert Corrosion," Technical Paper 76-5, Kenneth M. Jacobs, Maine DOT, August 1976.	The third of a sequence of reports on 43 aluminum alloy culverts and 3 aluminum structural plate culverts. The culverts had been exposed for 5 to 13 years in a variety of fresh water and salt water environments. The pH varied between 5.8 and 7.9 and resistivity was above 8250 $\Omega \cdot$cm. The observed deterioration was used to project a time to perforation well over 50 years.
"Accelerated Abrasion Test of Polymer Protective Coatings for Corrugated Metal Pipe," W.F. Crozier, J.R. Stoker, and E.F. Nordlin, January 1977.	This is a presentation of results from 30 accelerated abrasion tests. The test method seemed to be in accordance with the essentials of the proposed ASTM Standard with a rotating, close-ended steel drum with internally attached pipe samples. It was concluded that polymeric coatings do not necessarily possess abrasion resistance equal to that of hot-dipped asphalt coatings. It is concluded in this paper that accelerated laboratory tests are good indicators of relative abrasion resistance, but do not consider factors such as contamination of the coatings in production, damage of coatings in shipment and placement, and aging of the coatings.

"Performance Evaluation of Corrugated Metal Culverts in Georgia," Southeastern Corrugated Steel Pipe Association, 1977. ref. 3	This analysis is almost identical to the study by Brown and Kessler conducted in 1975 (reference 1). The difference with this study is that the field survey was of 251 culverts in Georgia rather than 153 culverts in Florida.
Durability of Drainage Pipe, Transportation Research Board, 1978.	This report is a synthesis of useful knowledge from many sources. The author discusses many of the major topics concerning CMP, including different coatings, theory and mechanisms of corrosion, protective measures, inspections, service-life estimation, and guidelines for durability.
Evaluation of Highway Culvert Coating Performance, W.T. Young, Federal Highway Administration, Report No. FHWA-RD-80-059, June 1980. ref. 10	In this study of CSP organic coatings, the authors reported less adhesion of a polymer precoat to Aluminized Type 2 than to galvanized ????.
Corrosion/Abrasion Performance of USS Nexon Culverts in Acidic Mine Wastewaters, U.S. Steel Corp., 1980.	This report discusses the results of the inspections of 11 installations of Nexon polymer-precoated CSP in Ohio and Pennsylvania (acidic mine-water environments). The locations varied in degree of abrasion and ranged from mildly corrosive to corrosive. The author presents data on the age, pH, bed load, and the general condition of the invert. This study concludes that Nexon culverts have lasted 10 times as long as plain galvanized steel in these acidic environments.
"Culvert Durability Study," Meacham, Hurd, and Shisler, Ohio, January 1982. ref. 19	This presents results of an extensive field investigation performed by the Ohio DOT with more than 1,600 culverts with various coatings. Many environmental conditions were considered. This study presents a mathematical model for predicting metal loss along with statistical predictions (based on field data) of culvert life expectancy.
Pipe Coating Study – Final Report, Indiana Department of Highways, September 1982. ref. 13	This is a discussion of parallel laboratory and field investigations. The laboratory investigation involved 29 pipe samples being exposed to an acid solution with a pH of 2. The field investigation involved evaluating performance of existing pipe installations in severe acidic environments. The data are mostly qualitative with many photographs. Resistivity and pH readings were recorded at various sites in southern Indiana.
"An Evaluation of Aluminized Steel Type 2 Culvert Coating," Richard F. Stratfull, January 14, 1982.	This is essentially a literature review done for Bethlehem Steel Corporation. It discusses failure mechanisms, protection mechanisms, expected soil-side and water-side performance, and six documents regarding Aluminized Type 2. There are no new data presented.
Durability of Drainage Structures, Final Report, Kenneth M. Jacobs, Transportation Research Board, 1982. ref. 11	This report is an evaluation of various culvert materials and/or coatings for durability, with emphasis directed primarily toward corrosion of the metal pipes and loss of aggregates in the reinforced concrete pipes. Estimated service lives were determined. A number of culverts were installed experimentally. More than 200 culverts were evaluated and the investigator concluded that pH had the major influence upon service life.
Culvert Committee Meeting Memorandum, Colorado Department of Highways, 1983. ref. 22	This report describes a method to estimate culvert life on the basis of sulphate, chloride, and pH measurements in the soil and water. These data establish a corrosion rating. Based on the corrosion rating, the engineer selects either CSP or concrete pipe.
Blac-Klad, Performance Report, Inland Steel, 1983.	This is a discussion of the general qualitative performance of polymer precoat pipe in various locations around the country.
"Corrosion Resistance of Aluminum Drainage Products," Sumerson for Kaiser Aluminum, January 1984.	This was a field study to evaluate the performance of aluminum drainage products. It included an evaluation of 67 Alclad pipe locations in California. The author discusses different methods for predicting service life.

"An Overview of Polymer Coatings for Corrugated Steel Pipe in New York," R.M. Pyskaldo and W.W. Renfrew, Transportation Research Board, 1984.	This study involved comparison between polymer-coated and asphalt-paved/polymer-coated corrugated steel pipe with asphalt-coated and paved pipe. Ten test sections were installed in nonaggressive sites and observations were made over a 7-year period. The only environmental data recorded were pH and all were close to neutral.
"Comprehensive Evaluation of Aluminized Steel Type 2 - Pipe Field Performance," G.E. Morris and L. Bednar, 1984.	Results of 30-year field tests of Aluminized Steel Type 2 and galvanized steel (in most cases the two were installed together) are presented. The results were tabulated from 54 tests sites of varying climatic conditions. The paper seems to be based on useful data, but the presentation of this data is lacking. The visual inspection of the sites is given in general terms and the criteria used are not defined.
Field Performance of Protective Linings for Concrete and Corrugated Steel Pipe Culverts, John Hurd, Ohio DOT, 1984.	This was an investigation into the durability of protective linings for concrete pipe and galvanized corrugated steel pipe used for culverts at corrosive and abrasive sites in Ohio. The performance of epoxy-coated concrete pipe, polymeric-coated corrugated steel pipe, and asbestos-bonded bituminous-coated-and-paved corrugated steel pipe were monitored for 10 years. The investigator concluded that all three materials provided satisfactory protection of pipe materials, except for the polymeric coating at abrasive sites. Data presented included various physical and environmental aspects.
Metal Loss Rate of Uncoated Steel and Aluminum Culverts in New York State, P.J. Bellair and J.P. Ewing, Transportation Research Board, 1984.	This is a description of a laboratory evaluation of three techniques to determine metal loss on 1-in coupons extracted from corrugated metal pipe. The field study involved the evaluation of 30 pipes to determine the sample size and sample location necessary to determine metal loss, and a statewide survey of steel and aluminum culverts in New York. The investigator concludes that a pin micrometer can be used to measure metal loss and that eight coupons extracted randomly along the "worst straight line" of the pipe will provide the needed accuracy for large-scale inspections. Average annual corrosion rates were based on the assumption of a linear time-corrosion relationship.
"Durability of Asphalt Coating and Paving on Corrugated Steel Culverts in New York," W.W. Renfrew, Transportation Research Board, 1984.	This paper describes a method developed to determine effectiveness of paving by measuring the longitudinal percentage of exposed metal. A field survey of 294 coated and paved pipes was performed. This study concluded that paving is ineffective in protecting any steel pipe after 30 years.
Durability of Polymer-Coated Corrugated Steel Pipe, Carl Hirsch, National Corrugated Steel Pipe Association, 1984.	This report is a tabulation of evaluation results of natural exposure testing of three classes of coating, including coal-tar-based polymer-coated, polymer-coated, and PVC Plastisol-coated CSP. The significant difference in test site bed-load conditions to which the various coatings were exposed bring question to the results of this study. Most of the sites with coal-tar-based coatings had significant bed loads while none of the sites with polymer-coated CSP had significant bed loads.
Galvalume Corrugated Steel Pipe: A Performance Summary, A.J. Stavros, 1984.	This is a laboratory and field investigation into the performance of Galvalume, galvanized, and Aluminized Type 2 coatings. Sheets with these coatings were subjected to laboratory testing designed to measure corrosion resistance to salt spray, cyclic standing water, full immersion, abrasion resistance, and asphalt adhesion. It was concluded that Galvalume sheet demonstrated the best relative performance. Results of the field tests substantiated laboratory results.

Bacterial Corrosion of Steel Pipe in Wisconsin, R. Patenaude, 1984.	Investigations into culvert corrosion in Wisconsin have indicated that anaerobic sulfate-reducing bacteria are a contributing factor in the corrosion of galvanized steel culverts. It was discovered by the investigator that at many sites where pH values were near neutral, corrosion rates far exceeded those predicted by the California Method, and at these sites the investigator has identified sulfate-reducing bacteriological activity.
"Culvert Durability – Where are We?," G.W. Ring, 1984.	This paper is a discussion of the progress to date developing guidelines for service-life predictions of culverts. The author presents graphical data summaries of the States' useful service lives for different types of culverts.
"Symposium on Durability of Culverts and Storm Drains," *Transportation Research Record 1001*, Transportation Research Board, National Research Council, 1984. ref. 15	This symposium contains 18 papers, most of which have been reviewed herein.
"Evaluation of Drainage Pipe by Field Experimentation and Supplemental Laboratory Experimentation," W. Temple, S. Cumbaa, and B. Gueho, March 1985. ref. 12	This study involves testing 16 different coatings (some are the same coating but different thickness) at 11 locations in Louisiana. The 11 locations represent 3 different corrosive environments. Environmental data were recorded at each site. Each coating is visually inspected at time intervals by more than one inspector. Comparisons are made against the California Chart.
"Analysis of Wyoming Highway Department Corrosion Resistance Criteria for Selection of Culvert Pipe," Thomas V. Edgar, January 1986.	This investigation involved a field survey of several highway reconstruction sites where high-corrosion-resistant pipe had been specified. After removal, the pipe was evaluated both qualitatively and quantitatively to determine if its condition indicates a need for a more corrosion-resistant pipe. Data were obtained on 13 pieces of pipe. This study analyzes inter-relations (i.e., weight loss versus resistivity, weight loss versus pH, weight loss versus percent soluble salts).
"Evaluation of CMP Invert Protection Products Under Different Bed Loads," California DOT, March 20, 1987. ref. 4	This is a laboratory investigation into the wear resistance of CMP coatings. A rotating drum with coated CMP samples bolted to the inside was used in the investigation. Galvanized steel was used as a standard to compare wear rates using a variety of aggregate grading. Much data are presented, including weight and thickness measurements at intervals. It was found that an elastomer over galvanized had the best wear resistance.
"Storm Sewer Pipe Material Technical Memorandum," Wright Engineers, Denver, CO, April 1987.	This discusses general factors involved with materials, design, installation, and durability of culverts. Consideration is placed on durability studies and guidelines from the FHWA, Colorado Department of Highways, Wyoming Highway Department, Oregon DOT, California DOT, and Los Angeles County Flood Control District.
"Performance Evaluation of Coated and Paved CSP Culverts," Clifton Ashford and J. Curtis Hayes, Dub Ross Co., April 1987.	This study presents a comparison of the abrasion and corrosion resistance of standard corrugated steel pipe to polymer-coated and coated/paved CSP culverts. The installation chosen is an area of high incidence of abrasion and corrosion. Useful field data were recorded for a few culverts, but the author only generally refers to these data.
Corrosion Potential Study, 56th Ave. Water Wells, Anchorage, AK, Harding, Lawson Associates, May 1987.	The scope of this report was to evaluate the corrosion risk along a roadway in Anchorage, Alaska, by collecting and analyzing groundwater samples and conducting an earth resistivity survey. Groundwater samples were compared against Langelier and Ryznar Indices to judge corrosivity.

"Metal Culvert Corrosion Study, Municipality of Anchorage, AK," Harding, Lawson Associates, May 1987.	This study provides an investigation of a corrosion problem in mid-town Anchorage, Alaska. Natural chemical and electrical properties were analyzed to determine the corrosive nature of the pipe environment. Groundwater samples were compared against Langelier and Ryznar Indices to judge corrosivity. This study uses the same methodology as the above study.
Accelerated Corrosion Test for Metal Drainage Pipes, James Garber and Jong Her Lin, FHWA/LA-87/206, June 1987.	This report describes development of an accelerated test to evaluate new coatings for steel culverts. The best method developed was a 3-day, high-pressure oxygen test at ambient temperature with a coupon placed inside a Baroid cell to elevate its potential.
"Corrosion Survey on Corrugated Steel Culvert Pipe for State of California," September 23, 1987.	This investigation involves 97 site inspections in California. Resistivity and pH readings were taken at all sites, along with 50 metal samples. Most of the culverts evaluated were 50+ years old. Culvert condition was noted and in some cases percent perforation was recorded. The author states that the California Method is too conservative and goes on to cite examples of pipes that have lasted well past their predicted service life.
Coatings for Corrugated Steel Pipe, R.M. Pyskadlo, New York, September 1987. ref. 9	This report summarizes site conditions of 71 culverts and eight 0.6- by 1.2-m (2- by 4-ft) test sections installed in culverts. A visual coating condition rating was used. Abrasion is the main focus of this study. Bed load is evaluated and recorded along with pipe slope and inlet slope.
"An Analysis of Visual Field Inspection Data of 900 Pipe-Arch Structures," Degler, Cowherd, and Hurd, January 1988.	This is a summary of data collected during the field inspection of 900 pipe-arch structures. Field evaluation was performed by 18 different inspectors. Data are statistically analyzed to determine dominant modes of structural failure and inter-relationships between structural failures and the variables of age, depth of cover, gauge of multiplate, and the graphical location.
Life-Cycle Cost for Drainage Structures, Potter, Technical Report GL-88-2, Department of the Army, COE, February 1988.	This provides a discussion on life-cycle cost analysis. Various coatings and environmental conditions are considered. Comparisons are made between the AISI and California life-prediction methods.
Evaluation of Corrugated Metal Pipe Arches, Vol. 1, Final Report, Degler, Perlea, and Cowherd, Bowser-Morner Associates, December 1988.	This report is a field evaluation of 890 pipe-arch structures. Dominant structural failure modes were determined using statistical analysis. From this analysis 50 structures were selected for comprehensive study and 10 were selected for subsurface analysis. The data presented concentrate mainly on CMP loading conditions with minor emphasis on corrosion.
Plain Galvanized Steel Drainage Pipe Durability Estimation With a Modified California Chart, Lawrence Bednar, Armco, Inc., Presented at Transportation Research Board Annual Meeting, January 1989.	This report addresses limitations of the California Chart with respect to pipe water-side corrosion. The author presents an estimation technique that takes scaling tendencies into account to make a modified California Chart. Field data from the United States and South America are presented as proof that the California Chart in its original form is generally over conservative and in some cases too liberal.
"An Evaluation of a Polymer Coating for Aluminum Coated Steel Pipe," Richard McKeon, NYDOT, August 1989.	This presents an evaluation of polymer bond to aluminum-coated (Type 2) steel. Testing of the bond was conducted on samples that were placed in a weatherometer.
"Aluminum-Coated Corrugated Steel Pipe Field Performance," John C. Potter, April 1990.	This article was a precursor to the 1991 Report (FHWA-FLP-91-006) (below). It contains essentially the same information.

Durability of Special Coatings for Corrugated Steel Pipe, J.C. Potter, I. Lewandowski, and D.W. White, Federal Highway Administration, Report No. FHWA-FLP-91-006, June 1991. ref. 6	This report includes a literature search and review with a limited field study updating research related to corrugated steel pipe and durability estimation. The scope of this investigation was to develop a procedure for durability estimation by using plain galvanized CSP as a baseline for comparison against CSP coated with various metallic and non-metallic coatings. Review of 80 references form the basis of this paper. Information collected in this study suggests that additional coatings can extend the service life of CSP to at least 50 years. Comparisons are made against the California Method. This paper seems to draw some valid conclusions based on field data and reviewed references.
Condition and Corrosion Survey on Corrugated Steel Storm Sewer and Culvert Pipe, Final Report, Corrpro Companies, March 1991.	This was an inspection and testing program to evaluate the long-term durability of corrugated steel pipe. Field evaluation was conducted of 122 sites across the United States (74 plain galvanized, 48 asphalt-coated). Data are presented for each site. Soil-side durability was the main focus of this study. IBM PC-compatible statistical model software was created from this study that predicts average service life where durability is limited by soil-side corrosion.
"Performance of Corrugated Steel Pipe Culverts Used in Low Water Stream Crossings in Eastern Oklahoma," Oklahoma DOT, August 1991. ref. 5	This study is a field evaluation of two Aluminum Type 2 sites and seven galvanized CMP sites. All sites have moderate to severe corrosion and/or erosion.
"Feasibility of Applying Cathodic Protection to Underground Corrugated Steel Pipe" J.D. Garber and J. H. Lin, University of Southwestern Louisiana, Transportation Research Board Annual Meeting, January 1992.	The Louisiana DOT study assessed the feasibility of applying cathodic protection to CSP both internally and externally to prevent corrosion. Field work involved the installation of 3-m (10-ft) test sections of CSP with various coatings. Current and potential measurements were taken during a 2-year exposure period.
"Updated Environmental Limits for Aluminized Steel Type 2 Pipe," L. Bednar, 72nd Annual Transportation Research Board Meeting, January 12, 1993.	This is a presentation of a modified Caltrans-type graph for use in life prediction of Aluminized Steel Type 2. This graph takes into consideration hardness and alkalinity along with pH and resistivity. The author does not present any data, but refers to data in other studies.
"Drainage Culvert Service Life Performance and Estimation," W. David Cerlianek and Rodney G. Powers, Florida DOT, April 1993. ref. 7	This is a large field study by the Florida DOT that concludes that Aluminum Type 2 lasts 2.9 times longer than the life prediction for galvanized CSP by the California Method.
Handbook of Steel Drainage and Highway Construction Products, American Iron and Steel Institute, Washington, DC, 1994. ref. 18	This is a comprehensive design manual. Of particular interest, it is one of a few sources to suggest a life extension of 25 years for a bituminous coating applied over galvanized CSP. It also provides AISI's method for durability prediction.

APPENDIX B. INSTRUCTIONS FOR DURABILITY ESTIMATION SOFTWARE

This durability estimation software has been developed to assist the engineer in culvert material selection. It uses computerized versions of existing methods to estimate the durability of a culvert. The durability prediction methods include those developed by Caltrans, AISI, NCSPA, and New York. The user should be familiar with the advantages and disadvantages of the various techniques. These methods are briefly described in an earlier section of this report.

This program will prompt the user for the various environmental conditions to which the culvert will be subjected. The program accepts the following inputs into alphanumeric strings for reference:

- Location.
- Run Date.
- User Name.
- Notes (any comments needed by user).

The following are used in the calculations:

- pH of the Water.
- pH of the Soil.
- Minimum Resistivity of the Water.
- Minimum Resistivity of the Soil.
- Abrasion Factor.

The program will then loop through a series of calculations based on the above variables. Recommendations and durability estimations based on the user's input of the environmental data will be the output. The output will be in tabular form and will be convenient for printing. The following are specific instructions for running the program:

1. Double-click the cmpdurab.exe icon or type "cmpdurab.exe" at the DOS command prompt to begin the program. A brief synopsis of the program is then displayed. After reading, press any key to start the program. You may also double-click on the icon to start the program.

2. The program will then prompt you for the data necessary to run the program. Please read the screens carefully before you enter any data. When entering data make sure not to use any commas and always press "enter" after every entry.

3. After you have entered all data the program will display the results of each method. Results from the American Iron & Steel Association, California Department of Transportation, and the Florida Department of Transportation are each displayed in three screens. First, a header displays the location, run date, user, pH, resistivity, abrasion factor, Cl%, and SO4%. The second screen contains the numerical results for each of the

methods. The last screen contains written recommendations to the user about the culvert. The three screens are only used in displaying the results on the computer. When printing, the results from each method will print on a separate page.

4.   Once you have viewed the reports, the program will prompt you if you want to run the New York Department of Transportation's method. Enter a "Y" if you do or an "N" if not.

"Y" - If you do, there will be a series of questions asked that should be answered by entering the displayed numbers. If you enter a number that is not displayed, the program will tell you that the values for the method do not meet the New York Department of Transportation test criteria.

"N" - If you enter an "N," the program will display the main menu.

5.   From the main menu you can print or display any of the methods the program has performed. The main menu is made up of two screens. Press any key to view the second screen. From the second screen you are given a choice of entering any of the numbered options or you can press enter to view the first screen of options again. Key in the number of the desired option and press enter.

6.   One of the features of this program is that it allows you to view any of the past culverts that you have entered. To do this, key in option number 15. This will cause the program to ask you to enter the number of the culvert you wish to view. The first culvert entered will be culvert #1, the second will be culvert #2, and so on. You can enter a maximum of 100 culverts.

After keying in the correct number a second menu will appear. It operates in the same manner as the main menu.

To go to the second menu for the program, enter option number 14.

The following examples may be helpful when familiarizing yourself with the program. Answer the questions with the underlined responses given and follow the instructions on the screen and the next few pages. Compare your results with those given for each example.

Example #1

Enter the location of the Culvert #1.
?<u>Maine</u>

Enter the Run Date.
?<u>today</u>

Enter User's Name.
?<u>your name</u>

Enter Notes.
?<u>none</u>

Enter pH of soil sample.
?<u>5.3</u>

Enter pH of water sample.
?<u>7.0</u>

Enter minimum resistivity of soil sample.
?<u>5000</u>

Enter minimum resistivity of water sample.
?<u>7000</u>

Enter the Abrasion Factor and/or velocity in fps.
0-5 Minimal, 6-8 Moderate, >8 Substantial Abrasion.
(Optional, if not entered default value will be 0.)
?<u>2</u>

Enter the Sulfate (SO4)% max of the soil sample.
(Optional, if not entered default value will be 0.)
?<u>1</u>

Enter the Sulfate (SO4) ppm max of the water sample.
(Optional, if not entered default value will be 0.)
?<u>2</u>

Enter the Chloride (Cl)% max of the soil sample.
(Optional, if not entered default value will be 0.)
?<u>1</u>

Enter the Chloride (Cl) ppm max of the water sample.
(Optional, if not entered default value will be 0.)
?<u>1</u>

(After completing the questions, your replies will be shown on the following screen. By hitting return you will be able to scroll through screens containing results pertaining to your data.)

Would you like to run the New York Culvert Design Criteria (Y/N)?

*   To run this method, the user should be familiar with the
        New York DOT Durability Index.

?<u>N</u>

(After answering the following question, a two-page menu will allow you to make a choice that supports your current needs.)

(In the case of this exercise, enter <u>17</u> to continue with the next example or <u>14</u> to exit the program.)

```
*************************************************************************
```

SOFTWARE FOR ESTIMATION OF DURABILITY FOR CULVERTS

```
*************************************************************************
```

Location of Culvert: Maine

Run Date: today

User: your name

```
*************************************************************************
```
Water pH – 7 , Soil pH – 5.3
Minimum Resistivity, ohm cm: Water – 7000 , Soil – 5000
Abrasion Factor – 2 , Soil SO4% – 1 , Water SO4 ppm – 2
Soil C1% – 1 , Water C1 ppm – 1
```
*************************************************************************
```


The NATIONAL CORRUGATD STEEL PIPE ASSOCIATION Recommends
Zinc or Aluminum-Zinc Alloy or Aluminum type II
With Protective Coatings.

PROTECTIVE COATINGS:

```
-------------------------------------------------------------------------
```

Polymer Coated	Bituminous Coated	Asbestos Bonded Coated
*Zinc AASHTO M 246	*Zinc AASHTO M 190	*Zinc Specification in
*Aluminum Type II M 246	*Aluminum type II M 190	preparation to
*Aluminum-Zinc Alloy M 246	*Aluminum-Zinc Alloy M 190	replace WWP405

Notes:
None

The California Method Estimated Perforation
Of Steel Culverts in Years

Gage	Thickness Inches	mm		Galvanized Water Data	Galvanized Soil Data
18	0.052	1.32		29	15
16	0.064	1.63		37	20
14	0.079	2.01		46	25
12	0.109	2.77		63	34
10	0.138	3.51		81	44
8	0.168	4.27		98	53

According to figure 854.3b of California Highway Design Manual 850
(dated 1992) aluminum culverts and aluminized coatings are not recommended
where the pH of the soil, backfill, and effluent is less than 5.5 or greater than 8.5.
Notes:
None

The American Iron and Steel Institute Estimated Average
Invert Life of Steel Culverts in Years

Gage	Thickness Inches	mm		Galvanized Water Data	Galvanized Soil Data
18	0.052	1.32		58	31
16	0.064	1.63		75	41
14	0.079	2.01		92	50
12	0.109	2.77		127	69
10	0.138	3.51		162	88
8	0.168	4.27		197	107

Notes:
None

```
*****************************************************************************
```
The Colorado Department of Highways
```
*****************************************************************************
```

In accordance with the Colorado Corrosion Resistant Culvert Specifications, it is recommended that the following corrosion resistant culverts be used.

- – Asbestos Bonded Corrugated Steel Pipe
- – Corrugated Aluminum Pipe
- – Reinforced Concrete Pipe of Type V Cement
- – Unreinforced Concrete Pipe of Type V Cement

Corrosion Condition Description: Severe   Corrosion Condition: Both Sides

---

Notes:
None

The Florida Department of Transportation
Estimated Service Life of
Galvanized Culverts in Years

Gage	Thickness		Galvanized Steel	
	Inches	mm	Water Data	Soil Data
16	0.064	1.63	36	19
14	0.079	2.01	47	25
12	0.109	2.77	65	35
10	0.138	3.51	83	45
8	0.168	4.27	101	55

The Florida Department of Transportation
Estimated Service Life of
Aluminized Culverts in Years

Gage	Thickness			Aluminized Steel	
	Inches	mm		Water Data	Soil Data
16	0.064	1.63		105	57
14	0.079	2.01		136	74
12	0.109	2.77		189	103
10	0.138	3.51		242	131
8	0.168	4.27		294	160

The Florida Department of Transportation
Estimated Service Life of
Alclad Aluminum Culverts in Years

Gage	Thickness			Alclad Aluminum	
	Inches	mm		Water Data	Soil Data
16	0.064	1.63		217	-5
14	0.079	2.01		282	-6
12	0.109	2.77		391	-9
10	0.138	3.51		499	-12
8	0.168	4.27		608	-14

Notes:
none

```
************************************************************************
```
THE NEW YORK DEPARTMENT OF TRANSPORTATION
```
************************************************************************
```
Data required for the NY Culvert Design Criteria was not entered.

Notes:
none

Example #2

Enter the location of the Culvert #2.
?Santiam Highway

Enter the Run Date.
?today

Enter User's Name.
?your name

Enter Notes.
?none

Enter pH of soil sample.
?6.2

Enter pH of water sample.
?6.9

Enter minimum resistivity of soil sample.
?1400

Enter minimum resistivity of water sample.
?6000

Enter the Abrasion Factor and/or velocity in fps.
0-5 Minimal, 6-8 Moderate, >8 Substantial Abrasion.
(Optional, if not entered default value will be 0.)
?3

Enter the Sulfate (SO4)% max of the soil sample.
(Optional, if not entered default value will be 0.)
?0

Enter the Sulfate (SO4) ppm max of the water sample.

(Optional, if not entered default value will be 0.)
?<u>0</u>

Enter the Chloride (Cl)% max of the soil sample.
(Optional, if not entered default value will be 0.)
?<u>0</u>

Enter the Chloride (Cl) ppm max of the water sample.
(Optional, if not entered default value will be 0.)
?<u>0</u>

(After completing the questions, your replies will be shown on the following screen. By hitting return you will be able to scroll through screens containing results pertaining to your data.)

Would you like to run the New York Culvert Design Criteria (Y/N)?

*   To run this method, the user should be familiar with the
         New York DOT Durability Index.

?<u>N</u>

(After answering the following question, a two-page menu will allow you to make a choice that supports your current needs.)

(In the case of this exercise, enter <u>17</u> to continue with the next example or <u>14</u> to exit the program.)

```
************************************************************************
            SOFTWARE FOR ESTIMATION OF DURABILITY FOR CULVERTS
************************************************************************
```

Location of Culvert: Santiam Highway

Run Date: today

User: your name

```
************************************************************************
```
Water pH – 6.9 , Soil pH – 6.2
Minimum Resistivity, ohm cm: Water – 6000 , Soil – 1400
Abrasion Factor – 3 , Soil SO4% – 0 , Water SO4 ppm – 0
Soil Cl% – 0 , Water Cl ppm – 0
```
************************************************************************
```

The National Corrugated Steel Pipe Association states that corrugated steel pipe with protective coatings have been used successfully below pH of 4 and above 12 and/or soil/water resistivities below 2,000 ohm-cm.
Special design practices are recommended for these installations.

Notes:
none

The California Method Estimated Perforation of Steel Culverts in Years

Gage	Thickness Inches	mm		Galvanized Water Data	Galvanized Soil Data
18	0.052	1.32		26	12
16	0.064	1.63		34	15
14	0.079	2.01		42	19
12	0.109	2.77		58	26
10	0.138	3.51		74	33
8	0.168	4.27		90	41

According to figure 854.3b of the California Highway Design Manual 850 (dated 1992) aluminum culverts and aluminized coatings are not recommended where the minimum resistivity of the soil, backfill, and effluent is less than 1500 ohm-cm.

The American Iron and Steel Institute Estimated Average

Invert Life of Steel Culverts in Years

Gage	Thickness			Galvanized Water Data	Galvanized Soil Data	
	Inches	mm				
18	0.052	1.32		53	24	
16	0.064	1.63		69	31	
14	0.079	2.01		85	38	
12	0.109	2.77		116		53
10	0.138	3.51		148		67
8	0.168	4.27		180		82

Notes:
None

****
The Colorado Department of Highways
****

In accordance with the Colorado Corrosion Resistant Culvert Specifications, it is recommended that the following corrosion resistant culverts be used.

_____

No special corrosion protection is recommended with the culvert.

_____

Notes:
none

_____

The Florida Department of Transportation
Estimated Services Life of
Galvanized Culverts in Years

Gage	Thickness			Galvanized Steel		
	Inches	mm		Water Data	Soil Data	
16	0.064	1.63		33	15	
14	0.079	2.01		43	19	
12	0.109	2.77		59	27	
10	0.138	3.51		76	34	
8	0.168	4.27		93	42	

_____

The Florida Department of Transportation
Estimated Services Life of
Aluminized Culverts in Years

Gage	Thickness		Aluminized Steel	
	Inches	mm	Water Data	Soil Data
16	0.064	1.63	96	43
14	0.079	2.01	125	56
12	0.109	2.77	173	78
10	0.138	3.51	221	100
8	0.168	4.27	269	122

The Florida Department of Transportation
Estimated Services Life of
Alclad Aluminum Culverts in Years

Gage	Thickness		Alclad Aluminum	
	Inches	mm	Water Data	Soil Data
16	0.064	1.63	211	159
14	0.079	2.01	274	207
12	0.109	2.77	380	287
10	0.138	3.51	486	366
8	0.168	4.27	591	446

Notes:
none

****
THE NEW YORK DEPARTMENT OF TRANSPORTATION
****

Data required for the NY Culvert Design Criteria was not entered.

Notes:
none

Example #3

Enter the location of the Culvert #3.

?<u>Maine</u>

Enter the Run Date.
?<u>today</u>

Enter User's Name.
?<u>your name</u>

Enter Notes.
?<u>none</u>

Enter pH of soil sample.
?<u>6.0</u>

Enter pH of water sample.
?<u>6.9</u>

Enter minimum resistivity of soil sample.
?<u>4000</u>

Enter minimum resistivity of water sample.
?<u>7000</u>

Enter the Abrasion Factor and/or velocity in fps.
0-5 Minimal, 6-8 Moderate, >8 Substantial Abrasion.
(Optional, if not entered default value will be 0.)
?<u>1</u>

Enter the Sulfate (SO4)% max of the soil sample.
(Optional, if not entered default value will be 0.)
?<u>2</u>

Enter the Sulfate (SO4) ppm max of the water sample.
(Optional, if not entered default value will be 0.)
?<u>25</u>

Enter the Chloride (Cl)% max of the soil sample.
(Optional, if not entered default value will be 0.)
?<u>3</u>

Enter the Chloride (Cl) ppm max of the water sample.
(Optional, if not entered default value will be 0.)
?<u>30</u>
(After completing the questions, your replies will be shown on the following screen. By hitting return you will be able to scroll through screens containing results pertaining to your data.)

Would you like to run the New York Culvert Design Criteria (Y/N)?

*   To run this method, the user should be familiar with the
New York DOT Durability Index.

?<u>Y</u>

Please enter the appropriate Rating for the following conditions.

SURFACE WATER CORROSIVENESS RATING

Relative Corrosiveness	Rating
Very Low	1
Low	2
Medium	3
High	7
Very High	9

Abnormally high corrosion rates related to unique local surface water conditions
(e.g., heavy agricultural, industrial waste, or mine tailings) require an adjusted rating or
the use of a different pipe material.

SURFACE WATER RATING = <u>2</u>

ABRASIVENESS RATING

A channel is potentially abrasive with a bed load of sand, gravel, and/or rocks whereas it is
generally non-abrasive with a bed load of silt, clay, or vegetation.

Abrasiveness Potential	Relative Abrasiveness	Rating
Non-Abrasive Bed Load	Low	1
Abrasive Bed Load, Gradient 2% or less	Low	1
Abrasive Bed Load, Gradient 2% to 4%	Medium	2
Abrasive Bed Load, Gradient 4% & up	High	5

ABRASIVE RATING = <u>1</u>

FLOW RATING

Flow Condition	General Pipe Usage	Rating

Highly Intermittent	Storm drains, slide drains, driveway pipes not subject to longstanding water.	1
Intermittent	Cross-culverts carrying streams or small flows that seasonally go dry.	2
Continuous	Cross-culverts carrying stream flows or any condition of longstanding water.	3

FLOW RATING = 2

SERVICE RATING

Service Category	Relative Rating	Rating
Slide Drains, Driveway Pipes	Low	1
Cross-Culverts	High	2

SERVICE RATING = 2

(After answering the following question, a two-page menu will allow you to make a choice that supports your current needs.)

(In the case of this exercise, enter 14 to exit the program.)

```
**********************************************************************
```
SOFTWARE FOR ESTIMATION OF DURABILITY FOR CULVERTS
```
**********************************************************************
```

Location of Culvert: Maine

Run Date: today

User: your name

```
**********************************************************************
```
Water pH – 6.9 , Soil pH – 6
Minimum Resistivity, ohm cm: Water – 7000 , Soil – 4000
Abrasion Factor – 1 , Soil SO4% – 2 , Water SO4 ppm – 25
Soil Cl% – 3, Water Cl ppm – 30
```
**********************************************************************
```
The National Corrugated Steel Pipe Association recommends Zinc or Aluminum-Zinc Alloy or Aluminum type II with No Protective Coatings.

PROTECTIVE COATINGS:

Polymer Coated	Bituminous Coated	Asbestos Bonded Coated
*Zinc AASHTO M 246	*Zinc AASHTO M 190	*Zinc Specification in
*Aluminum Type II M 246	*Aluminum type II M 190	preparation to
*Aluminum-Zinc Alloy M 246	*Aluminum-Zinc Alloy M 190	replace WWP405

Notes:
none

The California Method Estimated Perforation
of Steel Culverts in Years

Gage	Thickness Inches	mm		Galvanized Water Data	Galvanized Soil Data
18	0.052	1.32		27	17
16	0.064	1.63		35	22
14	0.079	2.01		43	27
12	0.109	2.77		60	38
10	0.138	3.51		76	48
8	0.168	4.27		93	58

According to figure 854.3b of California Highway Design Manual 850 (dated 1992), the following materials are acceptable for a 50-year maintenance free service life:
   * 0.060 in (16 ga) Aluminum pipe (non-abrasive conditions)
   * Aluminized Steel 2 gauges smaller than galvanized steel
   The American Iron and Steel Institute Estimated Average

Invert Life of Steel Culverts in Years

Gage	Thickness			Galvanized Water Data	Galvanized Soil Data
	Inches	mm			
18	0.052	1.32		54	34
16	0.064	1.63		71	45
14	0.079	2.01		87	55
12	0.109	2.77		120	76
10	0.138	3.51		153	96
8	0.168	4.27		186	117

Notes:
none

****

The Colorado Department of Highways

****

In accordance with the Colorado Corrosion Resistant Culvert Specifications, it is recommended that the following corrosion resistant culverts be used.

The data values for SO4, Cl, and pH are not within the given parameters for the Colorado Method's Specifications.
Please refer to Section 624 Corrosion Resistant Culverts from the Colorado Department of Highways.

Notes:
none

The Florida Department of Transportation
Estimated Services Life of
Galvanized Culverts in Years

Gage	Thickness			Galvanized Steel Water Data	Soil Data
	Inches	mm			
16	0.064	1.63		34	21
14	0.079	2.01		44	28
12	0.109	2.77		61	38
10	0.138	3.51		79	49
8	0.168	4.27		96	60

The Florida Department of Transportation
Estimated Services Life of
Aluminized Culverts in Years

Gage	Thickness		Aluminized Steel	
	Inches	mm	Water Data	Soil Data
16	0.064	1.63	99	62
14	0.079	2.01	129	81
12	0.109	2.77	179	113
10	0.138	3.51	229	144
8	0.168	4.27	279	175

The Florida Department of Transportation
Estimated Services Life of
Alclad Aluminum Culverts in Years

Gage	Thickness		Alclad Aluminum	
	Inches	mm	Water Data	Soil Data
16	0.064	1.63	271	-5
14	0.079	2.01	282	-6
12	0.109	2.77	391	-9
10	0.138	3.51	499	-11
8	0.168	4.27	608	-14

Notes:
none

****
THE NEW YORK DEPARTMENT OF TRANSPORTATION
****

In Accordance with NY Culvert Design Criteria it is recommended that
Unpaved Items can be used.

Notes:
none

APPENDIX C

The print version of Appendix C is a printout of the April 1998 CSP
Durability Guide as it appeared on the National Corrugated Steel Pipe
Association (NCSPA) website. Since the NCSPA Durability Guide is
subject to periodic updating, it is not feasible to maintain the most up-to-
date version on this website. For the latest version of the Durability
Guide, it is suggested that you access www.NCSPA.org/..

APPENDIX D
ANALYSIS OF ALL DATA GATHERED IN THIS STUDY

The analysis presented in this report centered on the evaluation of data collected on the Aluminized Type 2 coated culverts previously evaluated by Potter (see figure 7 in FHWA-FLP-91-006) plus five additional Aluminized Type 2 culverts in Maine. We evaluated the condition of 11 additional culverts in this study because of their convenient proximity to the Aluminized Type 2 culverts that we were evaluating. The data from all of these culverts is presented in Table 15.

The additional 11 culverts included three Aluminized Type 2 coated culverts. Figure 21 shows the regression analysis described in FHWA-FLP-91-006 with this data added to the 21 discussed in the body of the report (making a total of 24 Aluminized Type 2 coated culverts). The results are not significantly different than discussed in this report – Aluminized Type 2 has an apparent advantage of 8.6 over the California Method's prediction for galvanized pipe. Including the "outliers" reduces the ratio to 0.79. The mode of the extreme value analysis for the pipes exhibiting a low pitting mode is 28.4 μm (1.12 mils). The mode of the extreme value analysis for the pipes exhibiting a high pitting mode is 445 μm (17.5 mils).

In addition to reporting on the Aluminized Type 2 pipes, Table 15 also shows:

a) The single galvanized pipe in this study is behaving somewhat worse than the annualized California Method prediction. This location was part of the triple tandem installation in Maine. The deterioration of this galvanized section was considerably worse than that of the Galvalume or Aluminized Type 2 in the tandem installation.

b) The perforation level of the bituminous-coated pipes was near to that predicted by the California Method for plain galvanized. However, the bituminous-coated pipes all appeared to be subject to significant abrasion.

c) At the triple tandem site in Garland, Maine, the Galvalume section was showing less wear than the galvanized section and appeared comparable to the Aluminized Type 2 section.

d) The aluminum alloy pipe evaluated in this study was one of three that was located on the access road to the recreation area on the Natchez Trace Parkway. This site was not very aggressive. The pipe had lost 1.4 percent of its thickness after 39 percent of the California Method predicted life — resulting in a predicted life increase of roughly 28 times that predicted by the California Method.

Table 15. Summary of all field data collected in this study.
Natchez Trace(NT), Santiam Highway(SH), and Maine(ME) Sites

Culvert Location	Coating Type	Pipe Gage	Slope Deg.	Velocity fps	Bedload	Thickness - inches			Percent Perforation	Soil		Water		Calif Pred. Years	Actual Age	
						Crown	Invert	Waterline		pH	Resistivity	pH	Resistivity		Years	Percent Calif. Pred.
ME Dexter	Al Type 2	16	0	8	heavy	0.057	0.011	0.049	80.70%	6.7	5138	7.4	7389	30	16	53%
ME Garland	Al Type 2	14	3	1	minor	0.072	0.070	0.072	2.78%	6.9	8130	6.8	9147	44	10	23%
ME New Gloucester	Al Type 2	14	2	3	minor	0.072	0.071	0.073	1.39%	5.3		6.4	7426	38	16	42%
ME Orrington	Al Type 2	14	0	1	minor	0.076	0.070	0.071	7.89%	6.5	3609	6.9	6186	32	10	31%
ME Ripley	Al Type 2	12	1	2	minor	0.098	0.097	0.097	1.02%	5.4	5036	6.6	9677	36	16	44%
NT 309.5	Al Type 2	16	0	*	none	0.060	0.057	0.059	5.00%	6.5	2399	7.5	5217	21	14	67%
NT 310.2	Al Type 2	16	0	*	none	0.056	0.054	0.055	3.57%	6.9	3001	7.2	9259	28	14	50%
NT 310.0	Al Type 2	16	4	4	moderate	0.058	0.000	0.054	100.00%	6.8	4710	7.0	17241	30	14	47%
NT 310.1	Al Type 2	16	1	3	moderate	0.056	0.000	0.056	100.00%	5.8	2646	7.1	22556	17	14	82%
NT 310.6 North	Al Type 2	16	1	*	minor	0.057	0.054	0.056	5.26%	6.8	6148	7.2	5272	32	14	44%
NT 311.9 South	Al Type 2	16	3	*	minor	0.056	0.056	0.056	0.00%	7.6	1601	7.7	6061	38	14	37%
NT 312.4 East	Al Type 2	16	0	0	none	0.057	0.056	0.057	1.75%	7.0	2594	7.2	2609	29	14	48%
NT 312.4 Single	Al Type 2	16	0	*	none	0.057	0.057	dry	0.00%	7.0	2882	dry	dry	30	14	47%
SH 100+15	Al Type 2	16	5	4	none	0.058	0.056	0.058	3.45%	5.3	3258	7.5	23438	17	14	82%
SH 104+45 East	Al Type 2	10	3	4	none	0.126	0.124	0.123	1.59%	5.7	4631	7.2	19608	48	14	29%
SH 104+45 West	Al Type 2	10	3	5	moderate	0.130	0.129	0.128	0.77%	5.7	4631	7.2	19608	48	14	29%
SH 123+76	Al Type 2	16	3	2	none	0.056	0.055	0.055	1.79%	5.6	11152	6.6	18868	27	14	52%
SH 13+00	Al Type 2	16	3	1	minor	0.057	0.057	0.057	0.00%	6.7	3464	6.8	46875	26	14	54%
SH 18+20	Al Type 2	16	4	1	none	0.057	0.054	0.056	5.26%	6.5	10352	7.3	18987	32	14	44%
SH 38+12 East	Al Type 2	14	1	3	moderate	0.073	0.070	0.070	4.11%	5.4	2973	6.4	13700	21	14	67%
SH 38+12 West	Al Type 2	14	1	5	moderate	0.073	0.070	0.071	4.11%	5.4	2973	6.4	13700	21	14	67%
SH 44+50	Al Type 2	16	3	3	none	0.057	0.049	0.055	14.04%	5.4	4212	7.5	24000	19	14	74%
SH 90+38 East	Al Type 2	16	5	4	moderate	0.057	0.056	0.055	1.75%	6.3	3914	7.2	21898	23	14	61%
SH 90+38 West	Al Type 2	16	5	4	moderate	0.057	0.053	0.055	7.02%	6.3	3914	7.2	21898	23	14	61%
NT 312.4 West	AlClad	14	0	0	none	0.073	0.072	0.073	1.37%	7.0	2594	7.2	2609	37	14	38%
ME Garland	Galvalume	16	3	1	minor	0.058	0.056	0.058	3.45%	6.9	8130	6.8	9147	35	10	29%
ME Orrington	Galvalume	16	0	1	none	0.057	0.039	0.057	31.58%	6.5	3609	6.9	6186	25	10	40%
NT 311.9 North	Galvalume	16	3	*	minor	0.059	0.042	0.058	28.81%	7.6	1601	7.9	4000	38	14	37%
ME Garland	Galv. Zinc	12	3	1	minor	0.106	0.080	No Sample	24.53%	6.9	8130	6.8	9147	63	10	16%
NT 310.6 South	Bituminous	16	1	*	minor	0.058	0.023	0.055	60.34%	6.8	6148	7.1	4967	32	14	44%
SH 113+25	Bituminous	16	1	6	heavy	0.058	0.045	0.057	22.41%	6.4	8420	7.3	20408	30	14	47%
SH 119+20	Bituminous	16	5	7	heavy	0.059	0.044	0.057	25.42%	5.6	13275	7.8	19355	28	14	50%

* Indicates standing water (below peak of corrugation) present at time of inspection

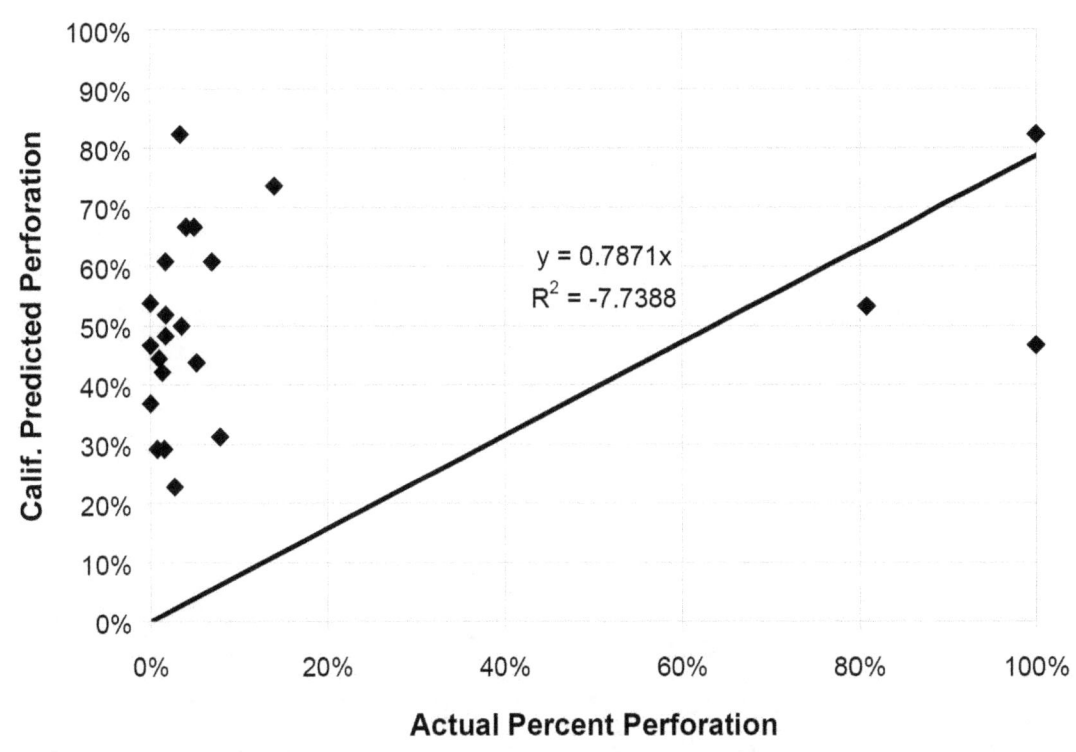

Figure 21. California Method curve prediction versus actual percent perforation, all Aluminized Type 2 in this study.

APPENDIX E.
ORIGINAL DATA AND ANALYSIS PRESENTED IN FHWA-FLP-91-006

This appendix presents data from Publication No. FHWA-FLP-91-006, *Durability of Special Coatings for Corrugated Steel Pipe*. This previous study investigated the performance of many of the pipes that were evaluated as part of the present work. Table 16 shows the Aluminized Type 2 data presented in FHWA-FLP-91-006, Table 4. Figure 22 is a recreation of figure 7 in FHWA-FLP-91-006. Both the table and the graph have been recreated in this report for convenience in referring to the previous work.

Table 16. Aluminized Type 2 Data from FHWA-FLP-91-006, Table 4

Location	Water		Soil		Thicknesses		
	pH	Minimum Resistivity (ohm-cm)	Ph	Minimum Resistivity (ohm-cm)	Original	Seven Year	Percent Perforation
NT 310.6	7.4	3,000	5.5	5,500	0.058	0.056	3.5
NT 311.9	7.1	2,000	7.5	3,100	0.058	0.056	3.5
NT 312.4	7.6	9,000	7.3	4,000	0.058	0.058	0.0
NT 312.4	7.6	9,000	7.3	4,000	0.058	0.057	1.7
SH 13+00	6.8	35,700	5.3	10,800	0.058	0.058	0.0
SH 18+20	6.4	46,500	4.2	9,400	0.058	0.058	0.0
SH W38+12	6.85	13,700	5.3	3,800	0.074	0.072	2.7
SH E38+12	6.85	13,700	5.3	3,800	0.074	0.072	2.7
SH 44+50	7.3	20,900	5.7	5,200	0.058	0.055	5.2
SH E90+38	6.7	15,500	5.5	2,600	0.057	0.056	1.8
SH 100+15	6.5	17,400	4.7	6,200	0.059	0.058	1.7
SH 104+45	6.8	20,000	5.7	6,600	0.128	0.125	2.3
SH 104+45	6.0	20,700	5.7	6,600	0.128	0.118	7.8
SH 123+76	6.3	15,400	5.5	4,900	0.057	0.057	0.0
NT 310.0[a]	7.6	9,000	2.5	290	0.058	0.000	100.0
NT 310.1[b]	8.6	29,000	2.5	290	0.058	0.009	84.5
[a]Fe = 6,575 mg/kg; Cu = 9.05 mg/kg; and SO$_4$ = 880 mg/kg [b]Fe = 6,380 mg/kg; Cu = 8.85 mg/kg; and SO$_4$ = 965 mg/kg							

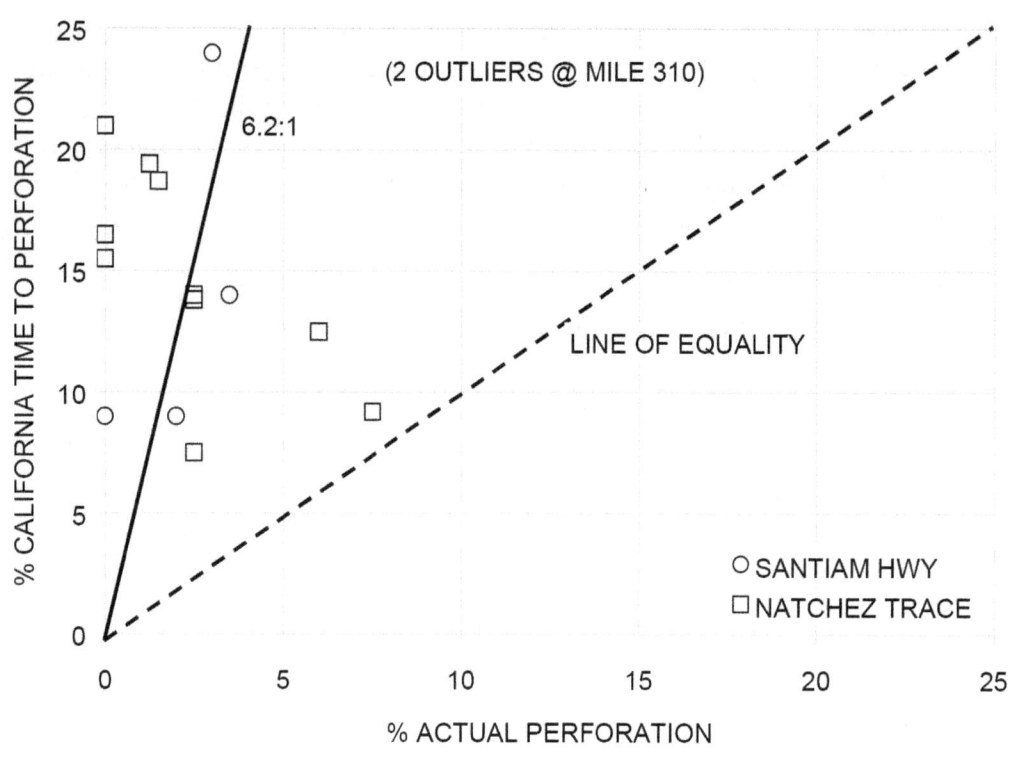

Figure 22. Aluminized type 2 performance as shown in figure 7 of FHWA-FLP-91-006.

REFERENCES

1. *Performance Evaluation of Corrugated Metal Culverts in Florida*, R.P. Brown, R.J. Kessler, Florida DOT, November, 1975.

2. *Durability of Metal Pipe Culverts*, State of Idaho, Department of Highways, 1965.

3. *Performance Evaluation of Corrugated Metal Culverts in Georgia*, Southeastern Corrugated Steel Pipe Association, 1977.

4. *Evaluation of CMP Invert Protection Products Under Different Bed Loads*, California DOT, March 20, 1987.

5. *Performance of Corrugated Steel Pipe Culverts Used in Low Water Stream Crossings in Eastern Oklahoma*, Oklahoma DOT, August 1991.

6. *Durability of Special Coatings for Corrugated Steel Pipe*, J.C. Potter, I. Lewandowski, and D.W. White, Federal Highway Administration, Report No. FHWA-FLP-91-006, June 1991.

7. *Drainage Culvert Service Life Performance and Estimation*, W. David Cerlianek and Rodney G. Powers, Florida DOT, April 1993.

8. *Highway Design Manual, Section 850-13*, California DOT.

9. *Coatings for Corrugated Steel Pipe*, R. Pyskadlo and J. Ewing, Engineering Research and Development Bureau, New York State DOT, September 1987.

10. *Evaluation of Highway Culvert Coating Performance*, W.T. Young, Federal Highway Administration, Report No. FHWA-RD-80-059, June 1980.

11. *Durability of Drainage Structures, Final Report*, Kenneth M. Jacobs, Maine DOT, Report No. BP-82(547), June 1982.

12. *Evaluation of Drainage Pipe by Field Experimentation and Supplemental Laboratory Experimentation*, W. Temple, S. Cumbaa, and B. Gueho, March 1985.

13. *Pipe Coating Study, Final Report*, Indiana Department of Highways, September 1982.

14. *Durability Design Method for Galvanized Steel Pipe in Iowa*, W.J. Malcom, 1968.

15. "Symposium on Durability of Culverts & Storm Drains,"*Transportation Research Record 1001*, Transportation Research Board, National Research Council, 1984

16. *Comparative Study of Coatings on Corrugrated Metal Culvert Pipe*, David K. Curtice and John E. Funnell, Southwest Research Institute, March 15, 1971.

17. *Field Test for Estimating the Service Life of Corrugated Metal Pipe Culverts*, State of California Department of Public Works, J.L. Beaton and R.F. Stratfull, January 1962.

18. *Handbook of Steel Drainage & Highway Construction Products*, American Iron and Steel Institute, Washington, DC, 1994.

19. *Culvert Durability Study*, Meacham, Hurd, and Shisler, Ohio DOT, Report No. ODOT/L&D/82-1, January 1982.

20. *Evaluation of Aluminum Alloy Pipe for Use in Utah's Highways*, Utah State Department of Highways, July 1973.

21. *Durability of Corrugated Steel Culvert in Oklahoma*, Oklahoma Department of Highways, C.J. Hayes, 1971.

22. "Culvert Committee Meeting Memorandum," Colorado Department of Highways, 1983.

23. *Evaluation Methodology for Corrugated Steel Pipe Coating/Invert Treatments,* National Corrugated Steel Pipe Association, Washington, DC, 1996.